a˜time

a˜tíme

***A**ll **T**here **I**s **M**atters **E**qually*

Kareline van der Burg

Frontier Publishing, Amsterdam, 2007

a~time All There Is Matters Equally ~ 2012 Maya Tzolkin Count ~ 13 Moon Zodiac Calendar ~ New Native Art

New Native Art: Kareline van der Burg
Jewellery design: Kareline van der Burg
Jewellery made in Bangkok with help of Eric Fabery de Jonge
Text: Kareline van der Burg
Photography: Anouk Jutta Heidweiler
Cover: photo of '9 heartbeats' by Anouk Jutta Heidweiler
Lay out & Illustrations: Kareline van der Burg
Typeset in Guardi, Papyrus and Square 721 BT
Printed in China

ISBN 978-1-931882-83-5

Frontier Publishing
P.O.Box 10681
1001 ER Amsterdam
the Netherlands
tel.: +31 (0)20-3309151
fax: +31 (0)20-3309150
e-mail: info@fsf.nl
www.frontierpublishing.nl

Adventures Unlimited Press
P.O. Box 74 , 303 Main Street
Kempton, Il 60946, USA
tel.: 815-253-6390
fax: 815-253-6300
e-mail: auphq@frontiernet.net
www.adventuresunlimitedpress.com

For Maximiliaan & Titus

To honour Gaia,

to embrace all breathtaking trees,

to cherish all colourful animals

and for all other wonderful children.

Contents

Introduction

This book is an attempt to make sense of what is the matter with us, people today, living somewhere on Earth. This project started on 9-11, a turning point for so many people. It made me think of America's karma, the loss of the Indian wisdom and the outright aggressiveness of America. It made me realise that the western way of living is no longer the way forward.

Mohawk haircuts and other Indian tokens became popular in artistic circles shortly after the 9-11 events. I started looking into Indian culture and their love and feel for the unspoiled nature around them, seemingly unparalleled in history. Indians viewed their interaction with the animals and plants as important lessons to help them understand their own true nature.

I taught my children, Max and Titus, that wherever the Indians used to live, nature flourished unrestrained. Not surprisingly, they turned into Indians right there, on the spot. The hunt for feathers, leather, and arrows to make them as real as possible, was the game. Pictures taken, ideas of better backgrounds lead to dragging the bag with Indian stuff everywhere and with lots of fun.

Meanwhile, I had switched to natural materials for my artistic work, to highlight the beauty of them: goatskin, feathers, wood and oil paint. I wanted round things to start. Roundness shapes the female body. The feminine characteristics of caring and nurturing needs emphasing to heal the Earth. The feminine side has been structurally suppressed during the ages. Our buildings lack roundness. An Indian tent, which is positioned to the stars, ~ a calendar by itself, ~ makes individuals interact as equals.

A round base to paint on. I paint to show a fundamental idea, feeling or energy as clear as possible and nothing else. I am painting things behind things. I look at a subject and think what is this about, what is the feeling here..? Then I pick up the paint, hold on to the feeling and sketch it out: bzzzzt, yes, bzzt, yes, bzzzzzzzzzt, oh no,.... cloth, piece of cloth to wipe it off. Off, that is better. Bzzzt, bzzt.... Yes that is it! (*Oils on skin)*
New Native Art revives old native values for a new world order close to nature's gifts.

Round as everything moves, turns and spirals. A reminder of the passing of time, the movement of our planet, the Moon and the Sun. I looked at our calendar and discovered its illogical way to divide a year. I inserted a 13 month calendar in this book, to give the Moon a feminine celestrial body, her rightful place back, and to remember her influence. The 13 months are named after the zodiac stars, to restore our relationship with the sky and regain our universal nature again. We are inseparably connected. Just like one cell can reveal the status of our whole body. Gaia, with all her life, represents the whole universe.

The ancient Maya engraved the date 2012 in many of their pyramids, 2,500 years ago, to point to a major astronomical event. They encouraged us to look at the middle of the Milky Way, the Hunab Ku. Their knowledge of our universe surprises our scientists even today. The whole meaning of the Mayan calendrical system is still not completely understood, but the expanding value to us is unquestionable. Therefore I have implemented the Tzolkin Count (their most holy calendar). I have looked at the glyphs engraved on the stones and tried to simplify them without losing the characteristic personality of each single universal energy. They can help us understand the ruling rhythm of our universe. I am totally aware that the given explanation about the Maya Calendar in this book is insufficient. Hopefully it will encourage you to read up on it, in one of the books mentioned as literature at the back. It is just the best I could do. From all people studying the Maya Calendar, I am most impressed by the work of John Major Jenkins: Maya Cosmogenesis 2012 and Galactic Alignment. When I visited Angkor Wat in Cambodia, our guide Mr. Ra, had shown Mr. Jenkins around and lively discussions about the Maya calendar and the meaning of Angkor Wat followed.

A round drum, to announce a new time is approaching: Gaia moves onto the next dimension and we with her. Time is quickening and new knowledge becomes available to grasp the nature of our universe. Our whole solar system is moving into the photon belt, by which more light is shining onto Earth. Gaia is cleansing itself by changing weather patterns. A drum, to feel the rhythms of our hearts, listen to it and act upon it. To teach our children new values to live one's life for. The beat to value Gaia as a living organism with a very wise soul.

Time to live up to your true nature. And nature will help you. If you immediately reject this, that's fine too. You will just not notice it happening. The power of thought is still highly underestimated. What I think, defines my future. Time for a reality check: 'What do I believe, expect and exclude from reality?' If your ideas about reality are broad, that's what your life is most likely to be. Anything you rule out, will pass you by, literally. Crop circles, orbs, rediscovery of ancient knowledge, rediscovery of the true power of humans, the infinite wisdom of nature as a small part of our universe, changing DNA, are just a few eye-opening events happening now. What a privilege to live in these highly interesting times!

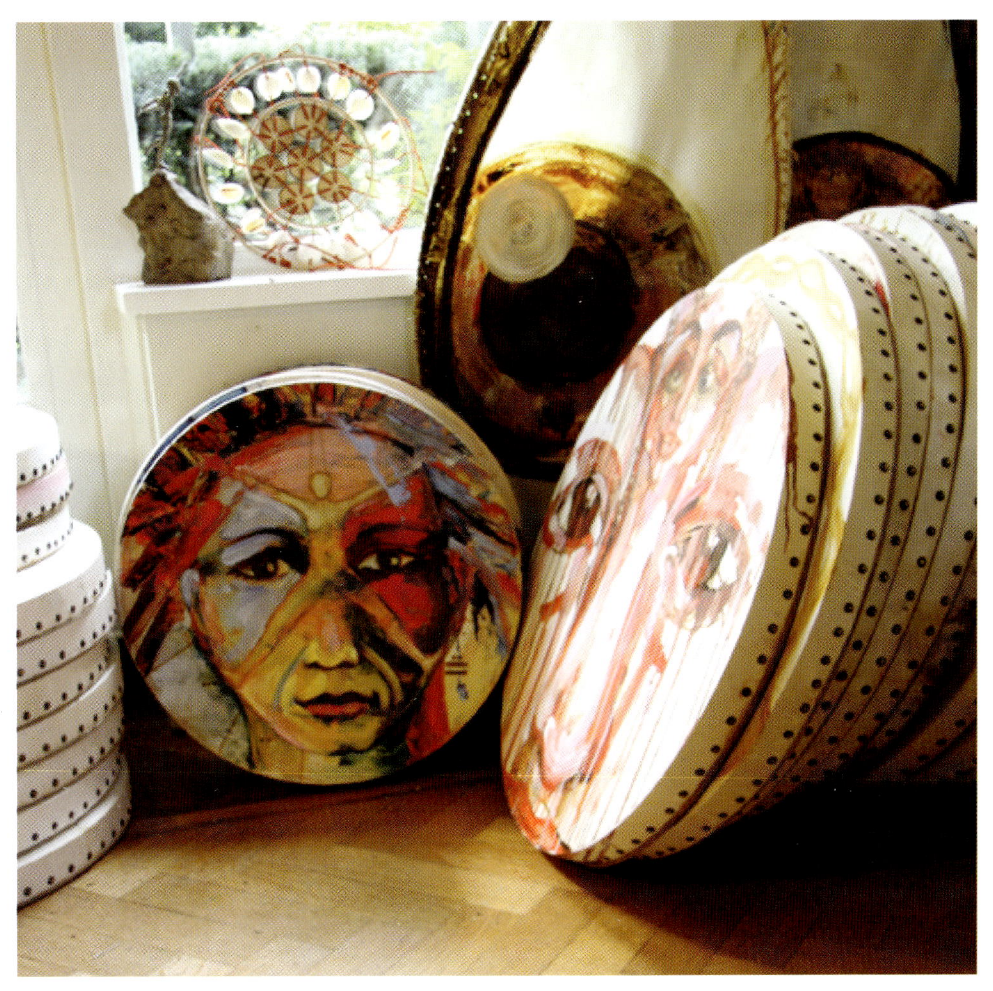

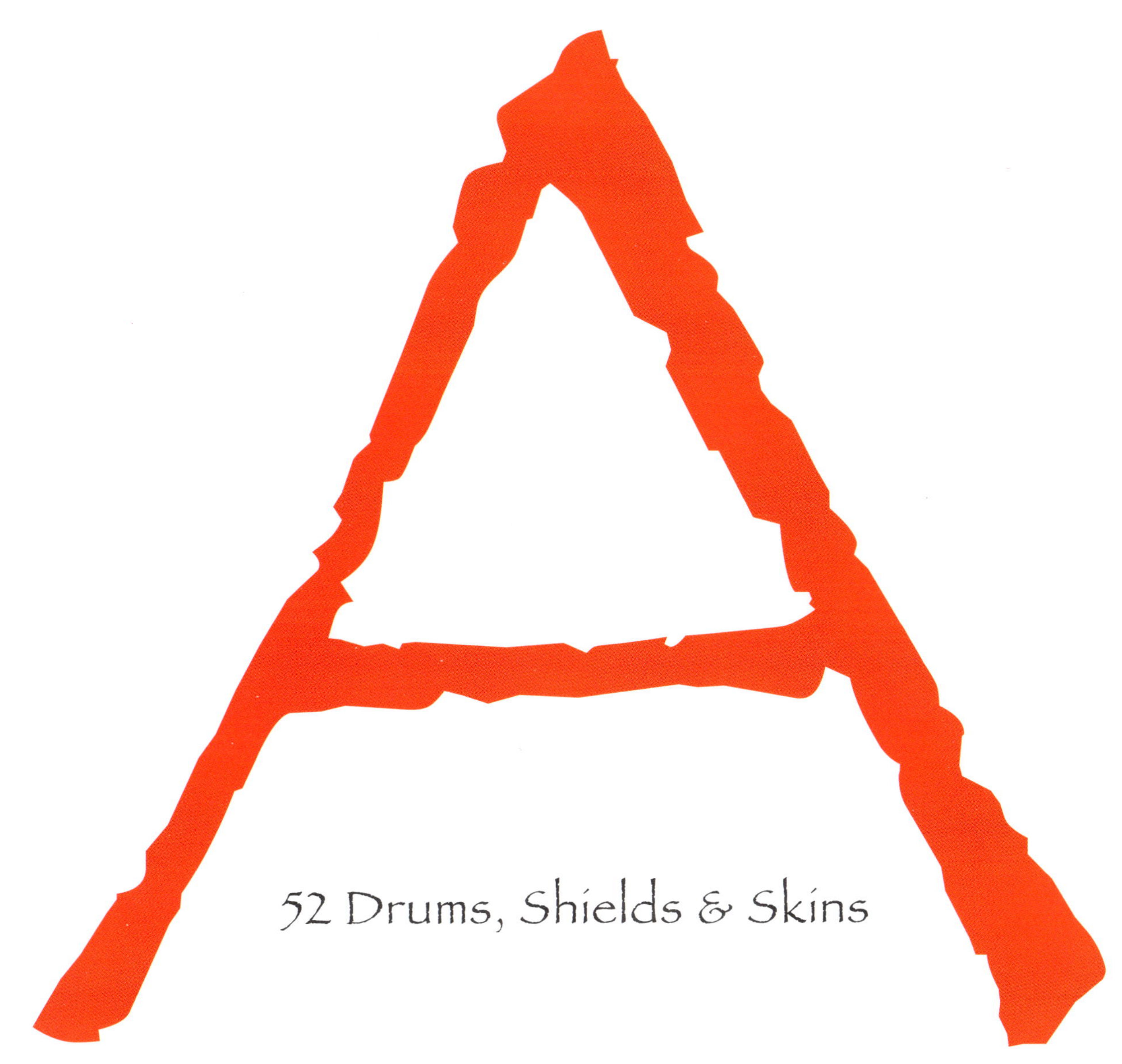
52 Drums, Shields & Skins

Matter

A~TIME

All There Is Matters, Equally

....Maybe not to you....

Some nonsense you cannot care less about..,

happens to be the reason to live, to me.

Everything around has value

or it would not exist.

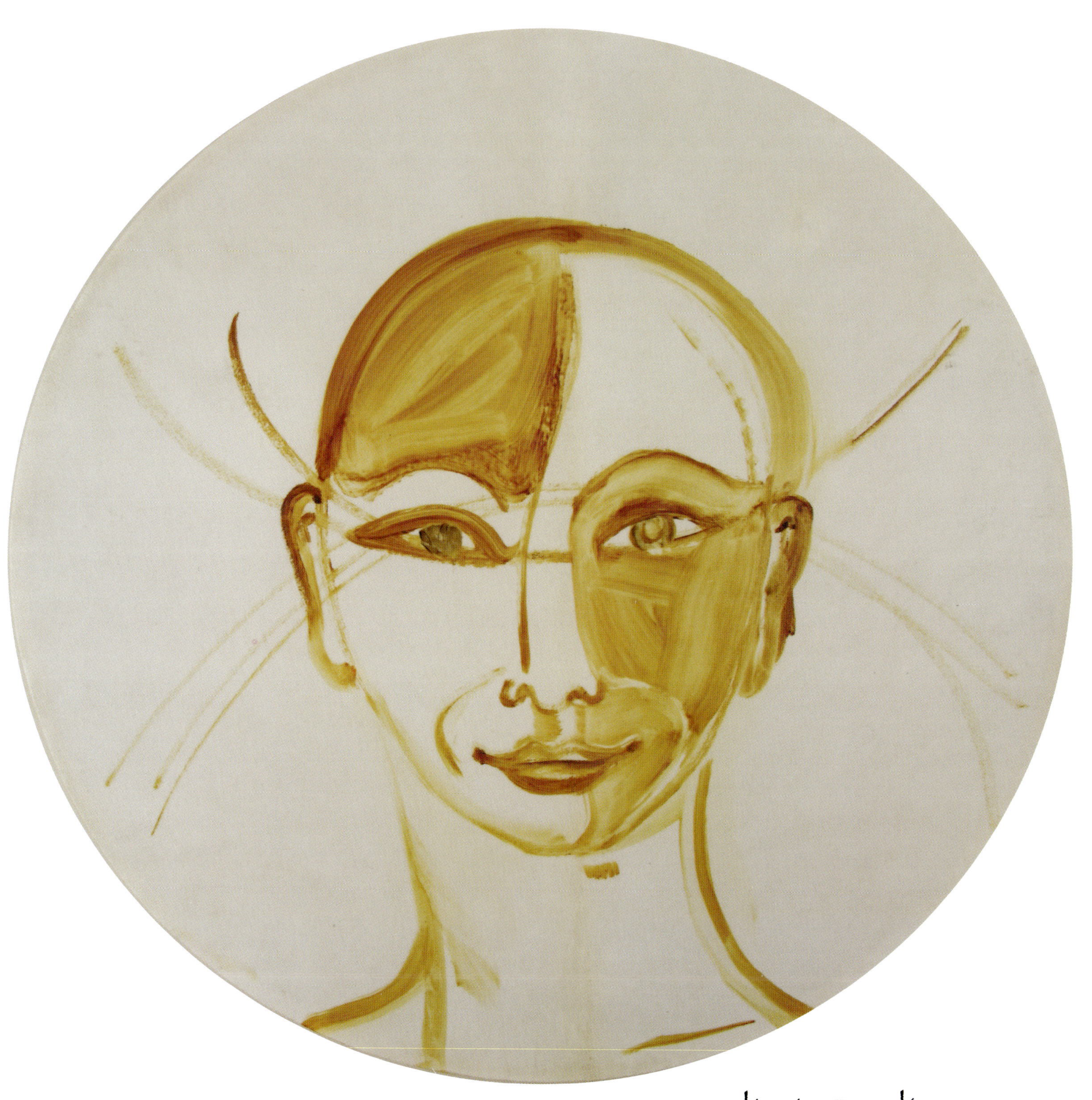

Dualistic Reality
Oils on Goatskin Drum ø 65 cm

Matter

All matter has a soul

THAT IS HOW THINGS START, TO BEGIN WITH.

When the soul leaves,

all that matters is lost

and the thing starts falling apart

leaving nothing behind.

Maybe a faint feeling...

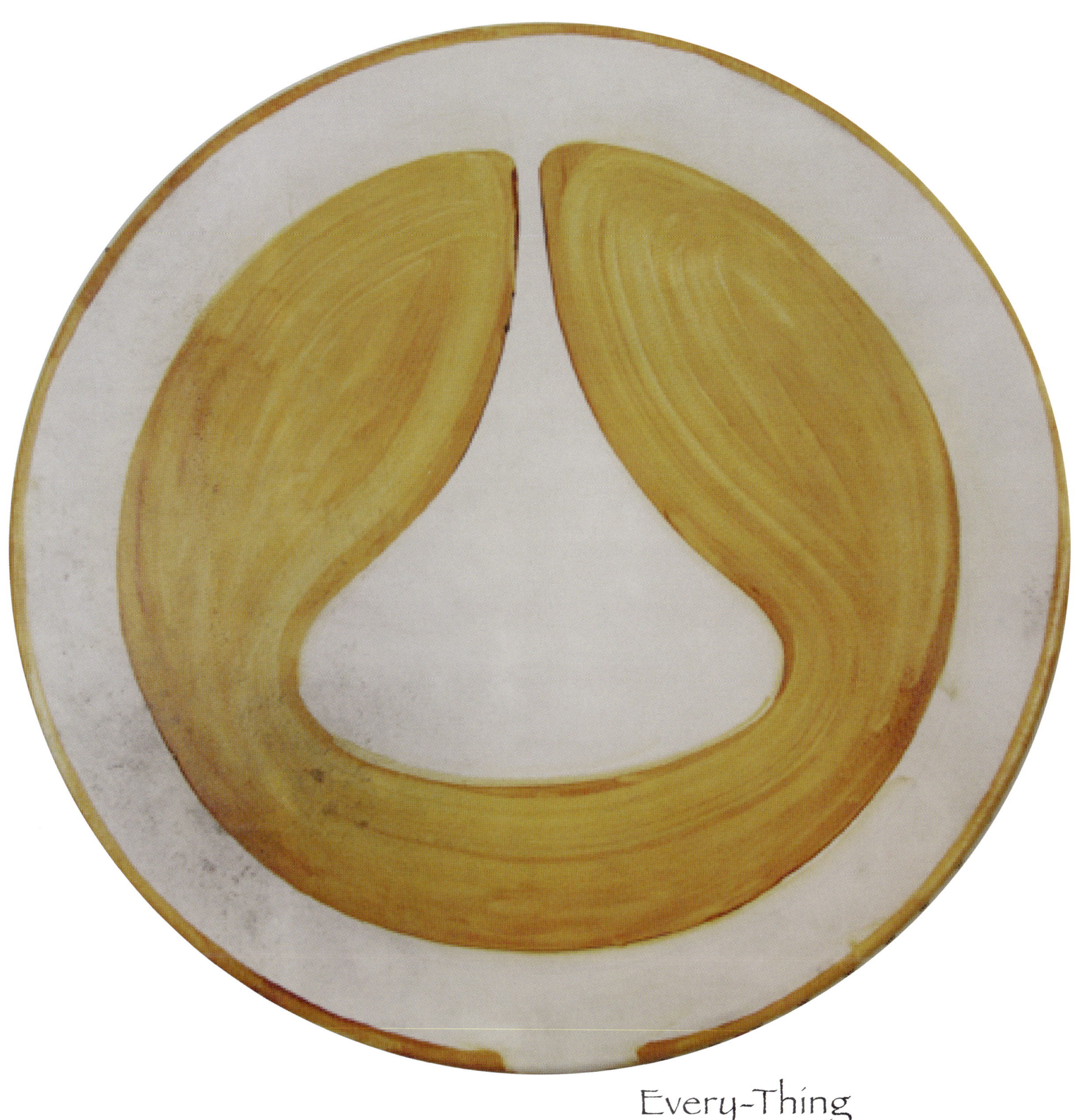

Every-Thing
Oils on Goatskin Drum ø 50 cm

Matter

What matters most,

..is connected to you..

..it comes to you..

..it flies around you..

Not too far, within reach,

but untouchable.

It is the soul stuff.

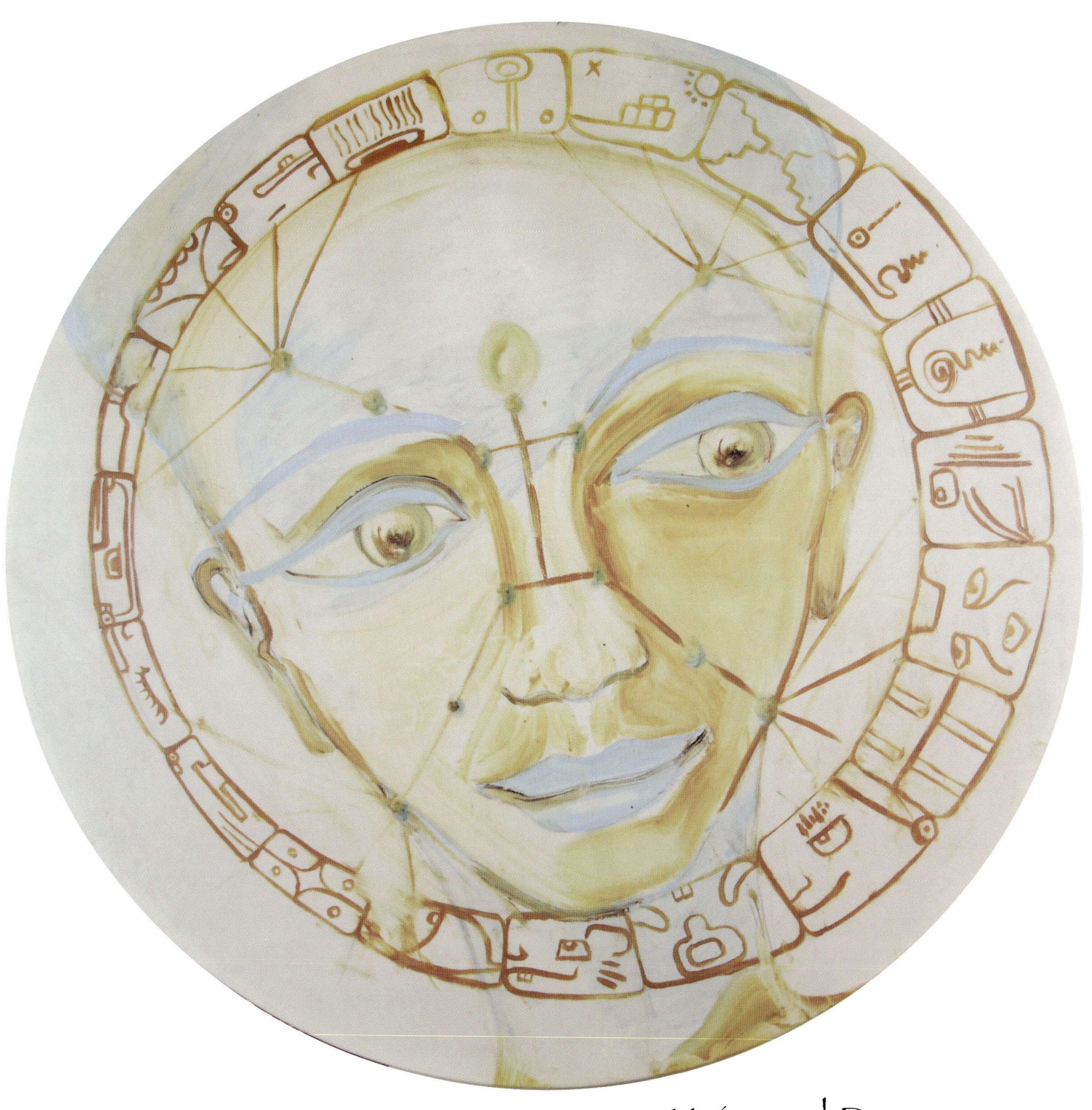

Universal Powers

Oils on Goatskin Drum ø 70 cm

· Matter

THEREFORE, NOTHING IS FOREVER

It's just passing by,

having a life

to live and die,

like everyone else around.

Only soul mates are friends for eternity

No matter what!

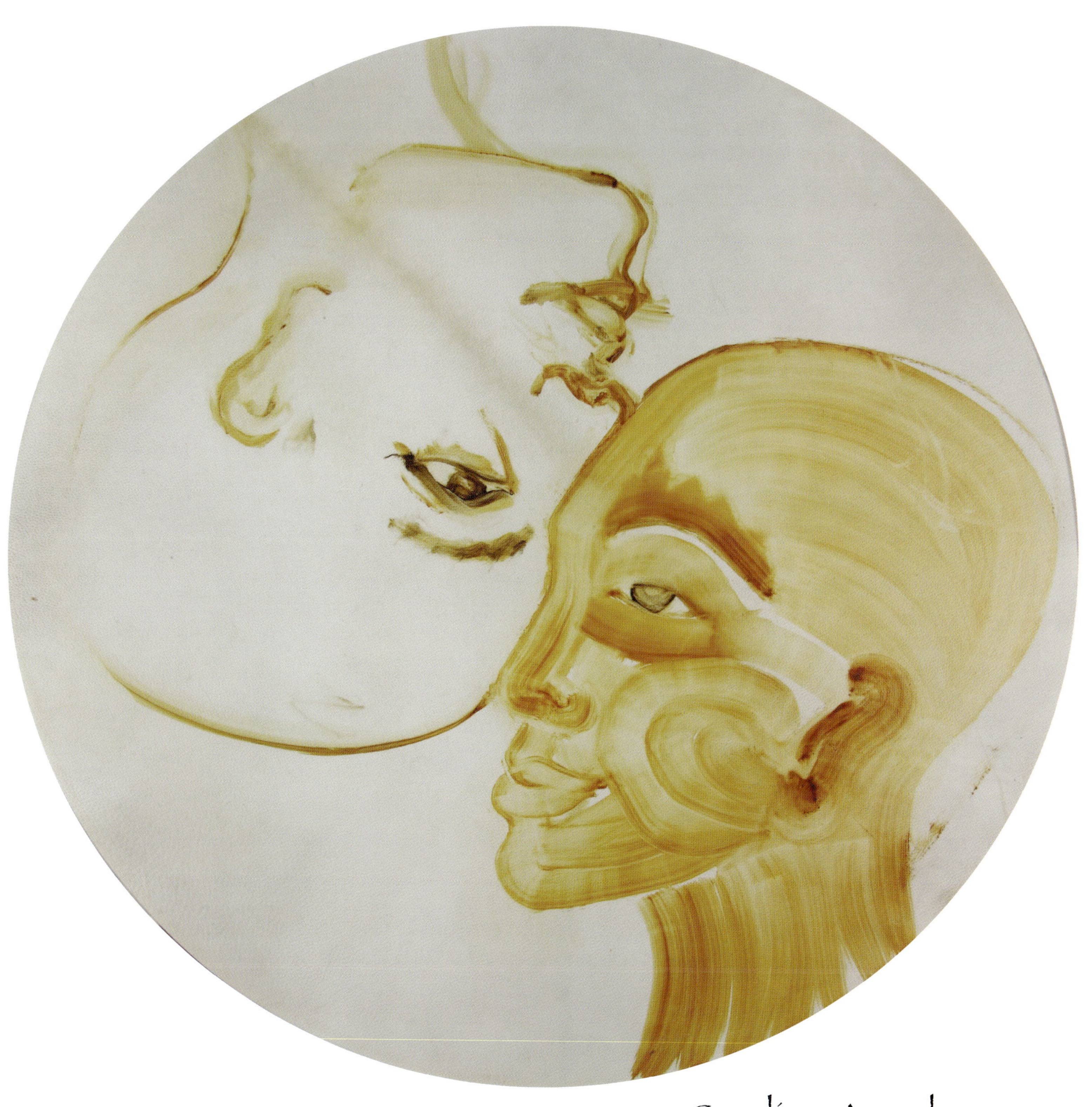

Gardian Angel
Oils on Goatskin Drum ø 65 cm

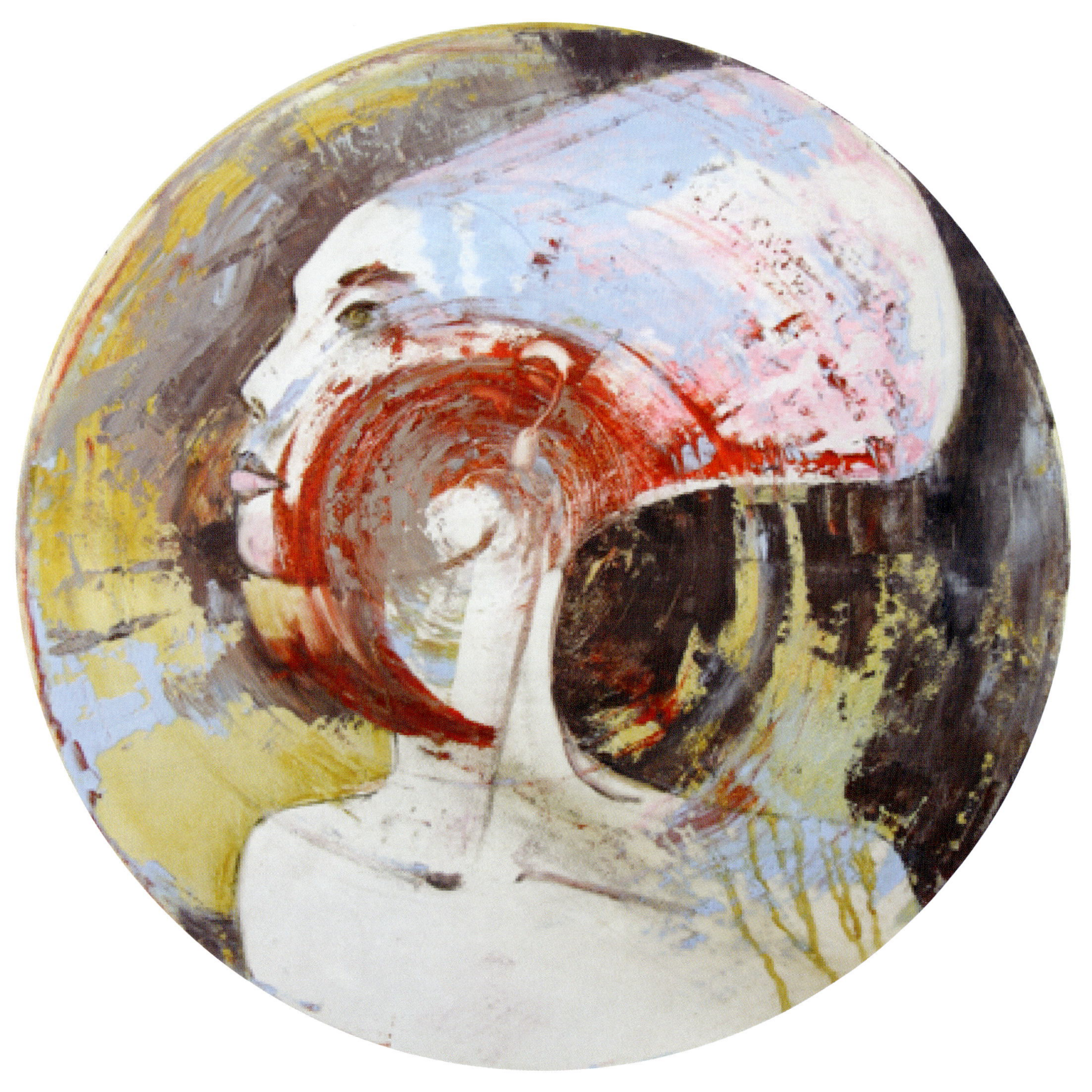

Brain Power
Oils on goatskin drum ø50cm

•• Time

TIME MATTERS

ALL THE TIME.

Time divides things from nothings.

To be or not to be,

it's all a matter of time.

Matter changes over time,

time changes our understanding of all matters.

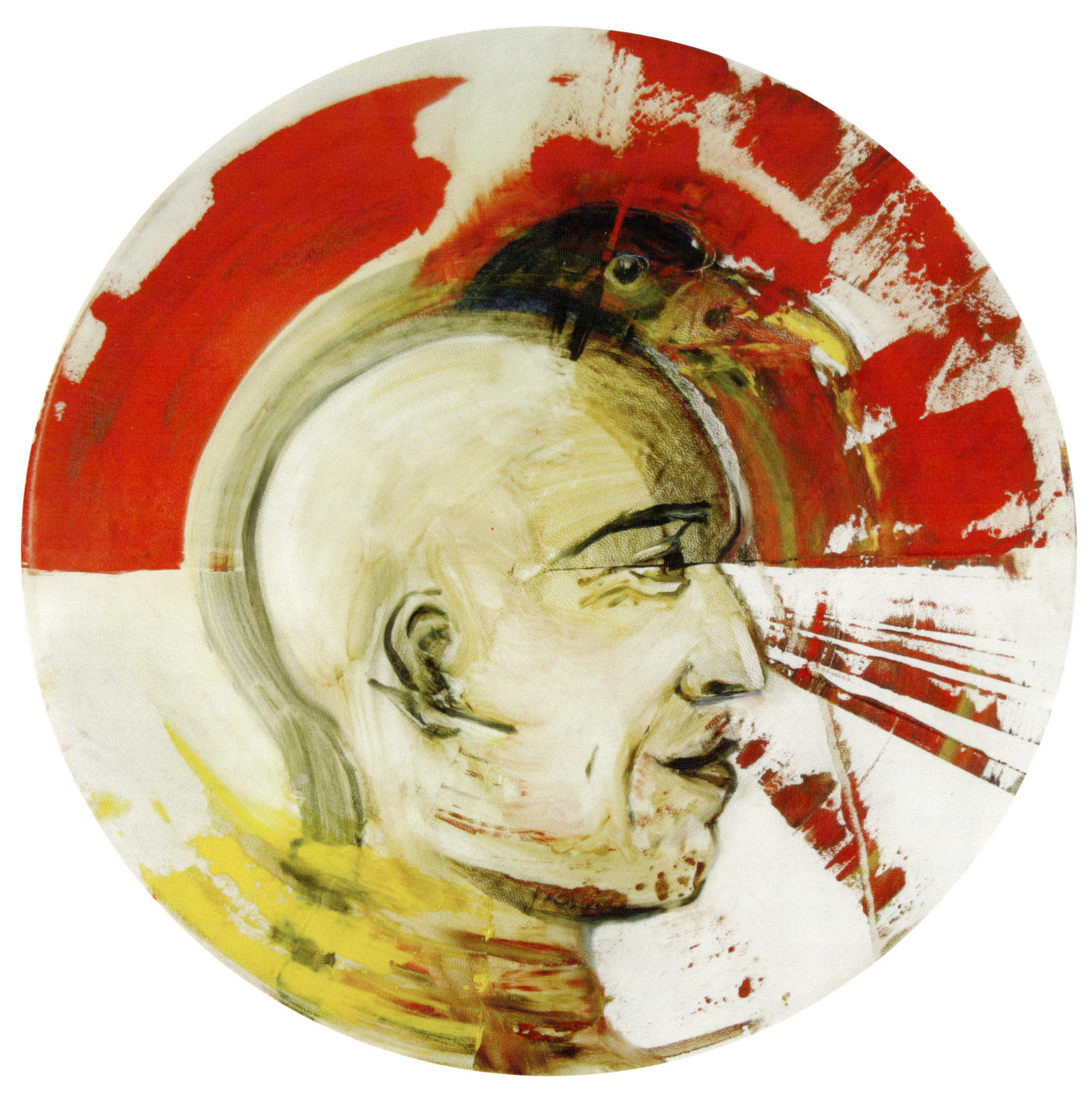

Birds' Eye View
Oils on goatskin drum ø50 cm

Time

WHAT YOU THINK,

DEFINES THE TIME WE LIVE IN.

To change the time we live by,

will change our way of thinking.

The only hope for Earth

at present, in human development

is just that.

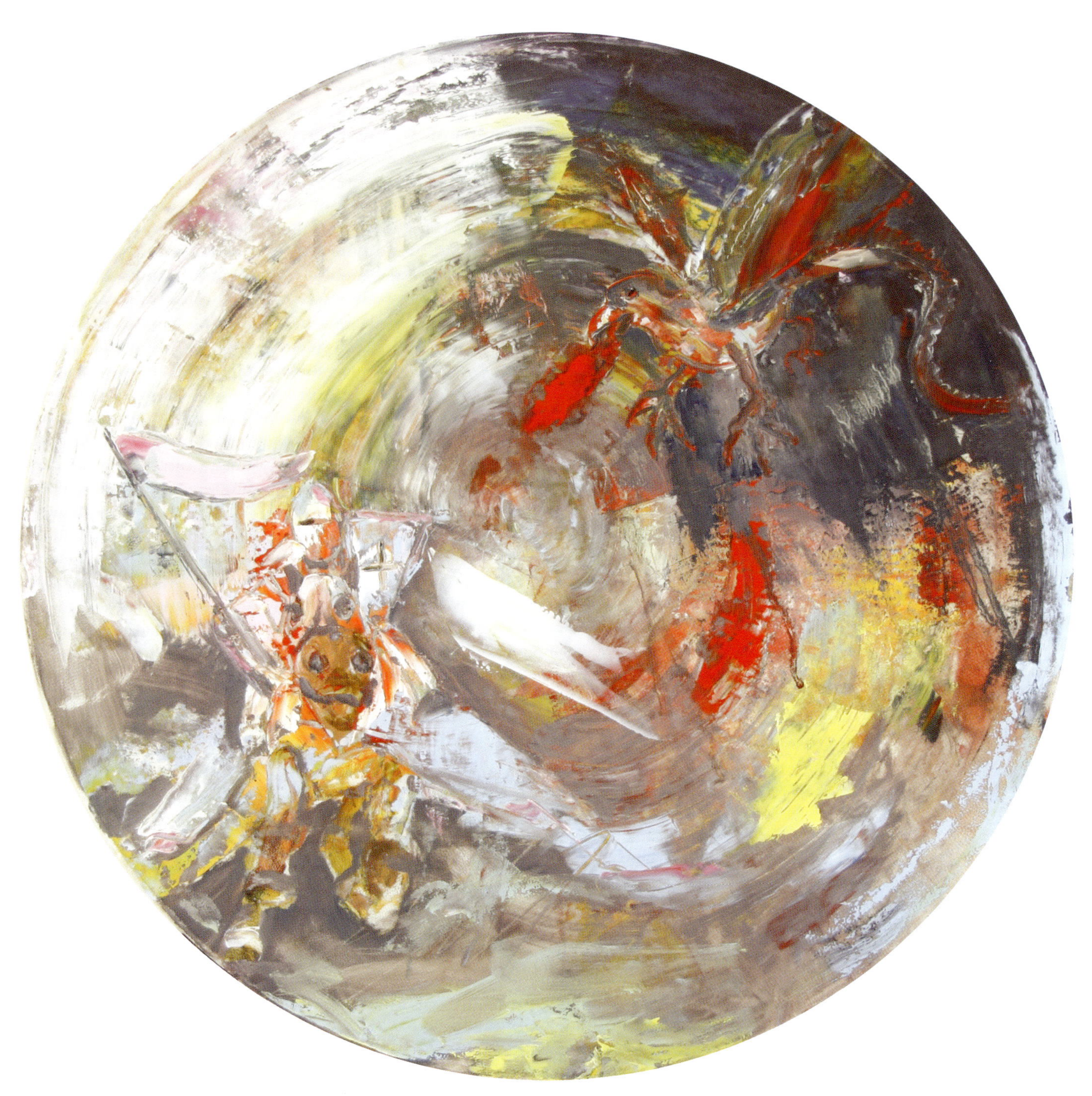

Knights Templers
Oils on goatskin drum ø50 cm

·· Time

The Gregorian calendar we are used to

is very chaotic, lacking order and logic.

So we think everything happens at random,

just what Julius Caesar hoped for.

He introduced this calendar

to control masses of people

by forcing an unnatural rhythm upon them.

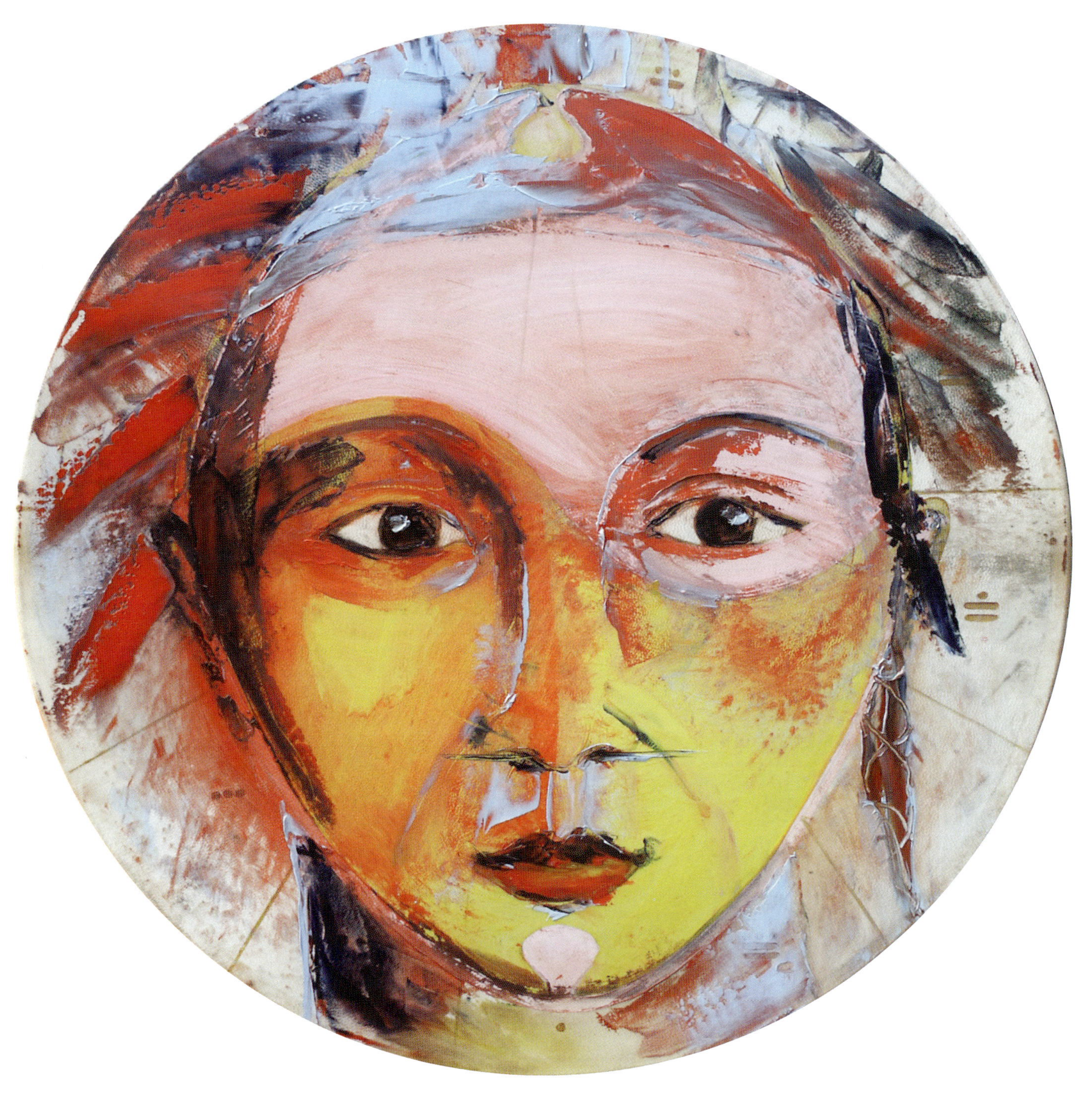

Listening to Nature
Oils on Goatskin Drum ø 50 cm

Time

A calendar of 13 months

can be a useful tool to help us

get in tune with the natural forces again.

It becomes 13 full Moon,

women have their cycle 13 times a year

The Egyptians, the Celts and the Mayan

all had calendars based on our natural rhythm,

To celebrate the beauty of balance.

... Money

Rich countries today live by poor ideology,

compared to civilisations in history.

The side effect of our advancements

are so toxicating that the American dream

starts to become a nightmare.

And the so called global market,

is closed to half the countries in the world.

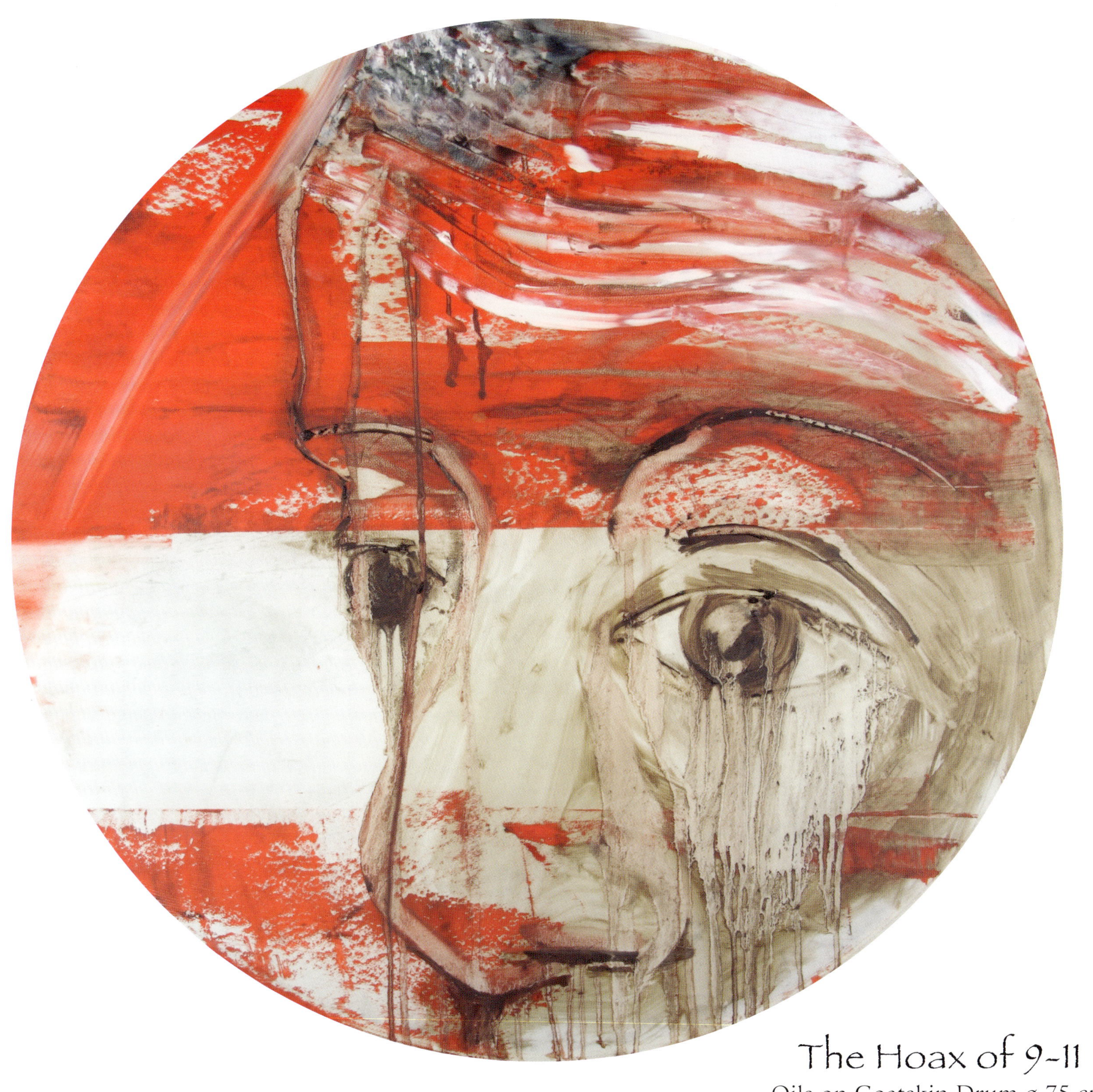

The Hoax of 9-11

Oils on Goatskin Drum ø 75 cm

… Money

We think the Earth is there, just for us

to use as we see fit.

With our superior attitude,

we ruined most nature,

many types of plants are extinct already,

many animal species endangered,

the seas are polluted and the soil get poorer by the minute.

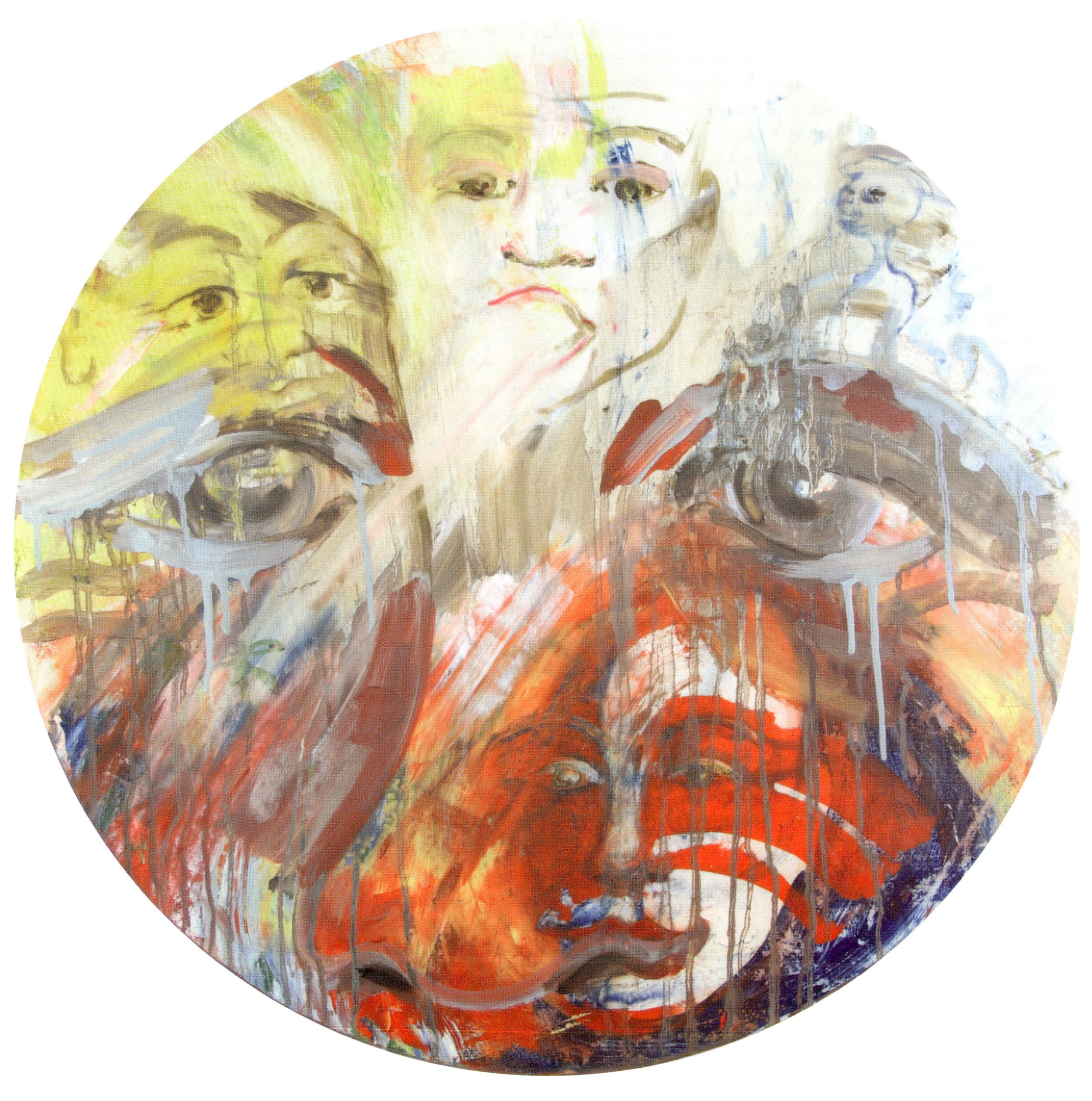

Sad Nature

Oils on Goatskin Drum ø 75 cm

... Money

Why?

We need the natural resources,

to make products to fulfill our wishes

We have not dealt with the negative sides of manufacturing

So, we have polluted almost everything around us

....Getting very sick of it now....

Still it has not brought us the advertised contentment

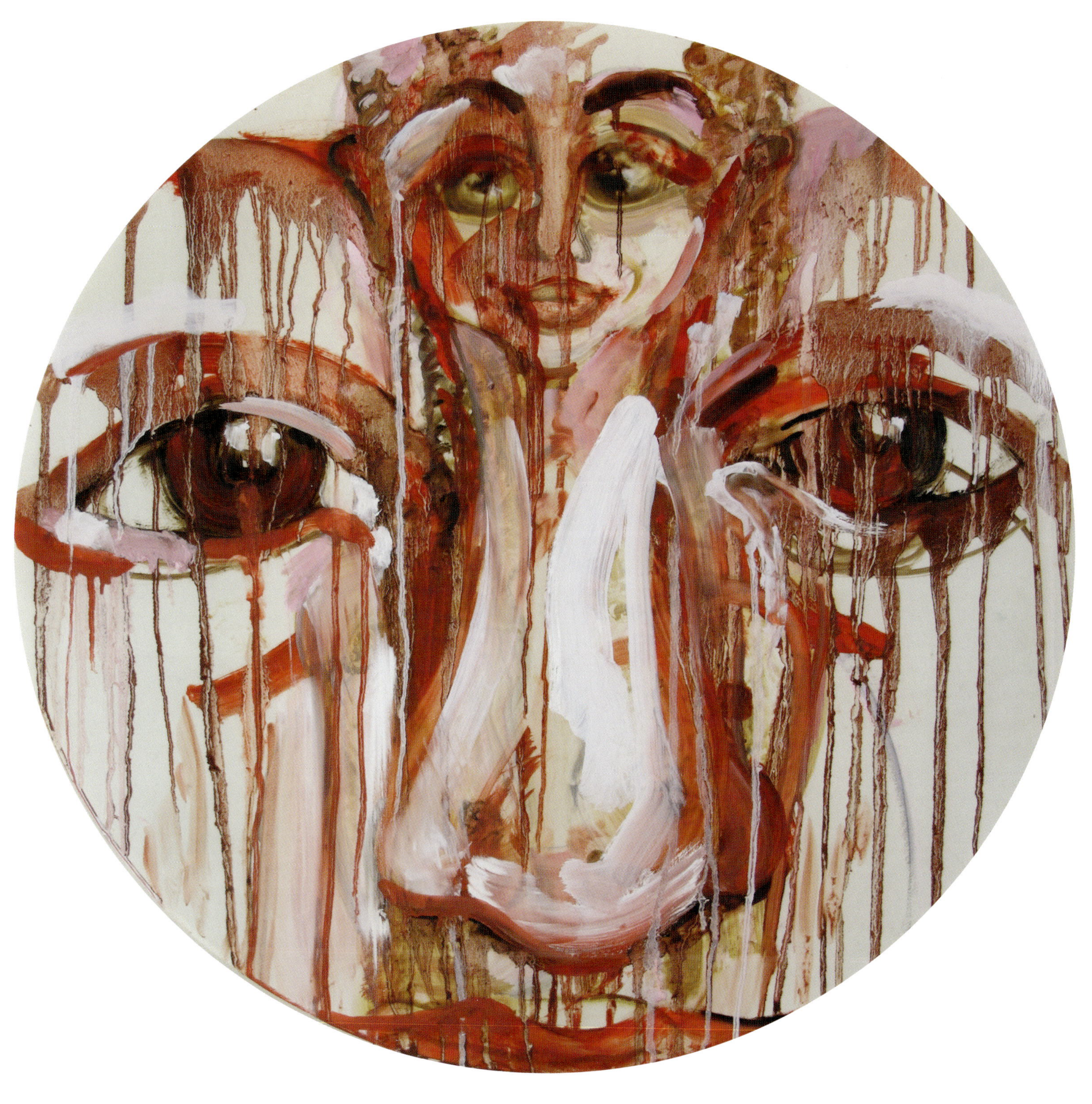

Unsatisfied Expectations
Oils on Goatskin Drum ø 75 cm

... Money

Why do we carry on?

Because there is money to be made

THE NEGATIVE POWER OF THE INTEREST DRIVEN MONEY SYSTEM

OUTWEIGHS THE POSITIVE BY FAR

It fuels the lower human instincts,

like fear and greed

and it values the necessities of life falsely.

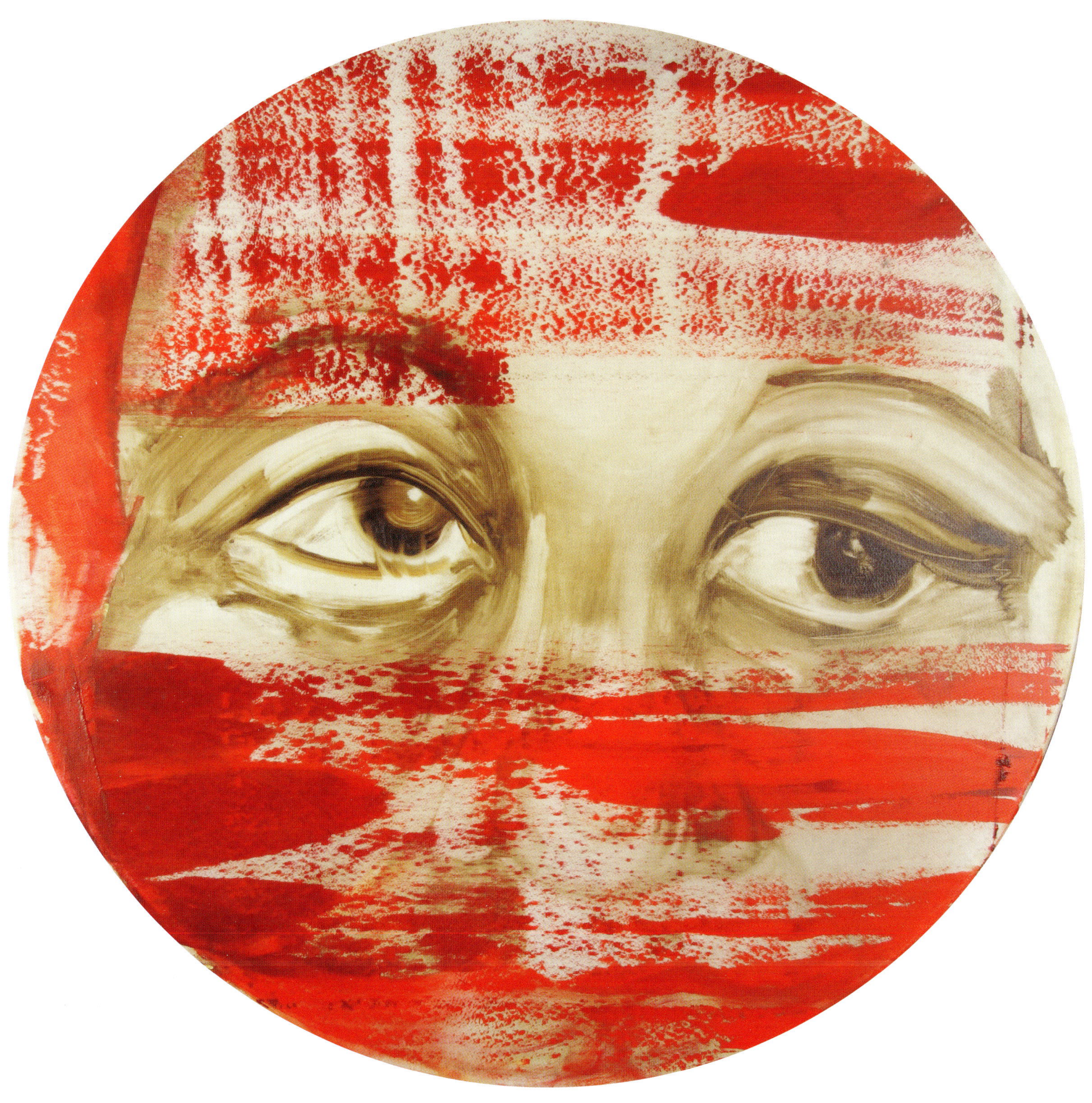

Doubt?

Oils on Goatskin Drum ø 75 cm

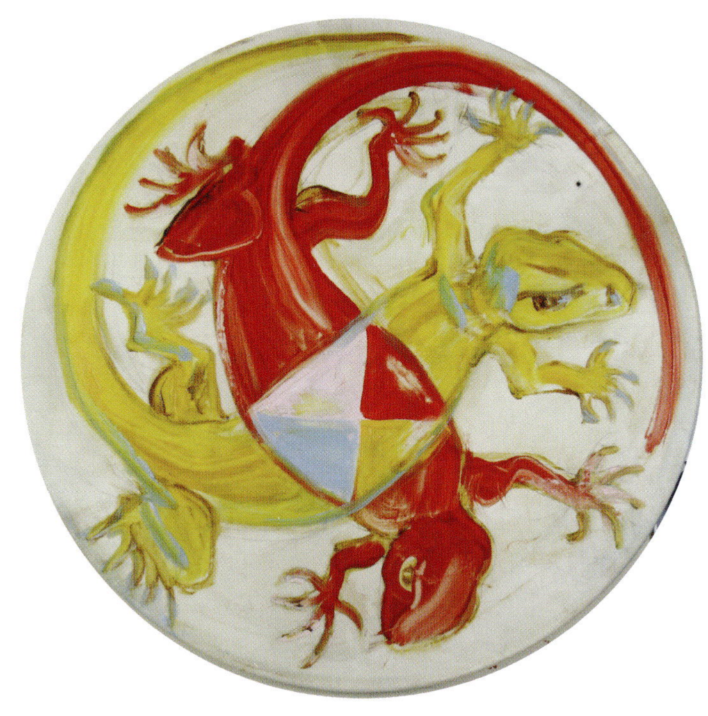

We All Feed of the Same Planet

Oils on Goatskin Drum ø 50 cm

.... Western Ways

We, in the west, adore our free democracy.

But how much do we actually have to say in matters?

MAJOR CHANNELS OF COMMUNICATION IN SOCIETY

ARE CONTROLLED CONCERNING THE KEY POLITICS

Free choice is realised by global providers

dominating the world, while profiting hugely

from poor countries.

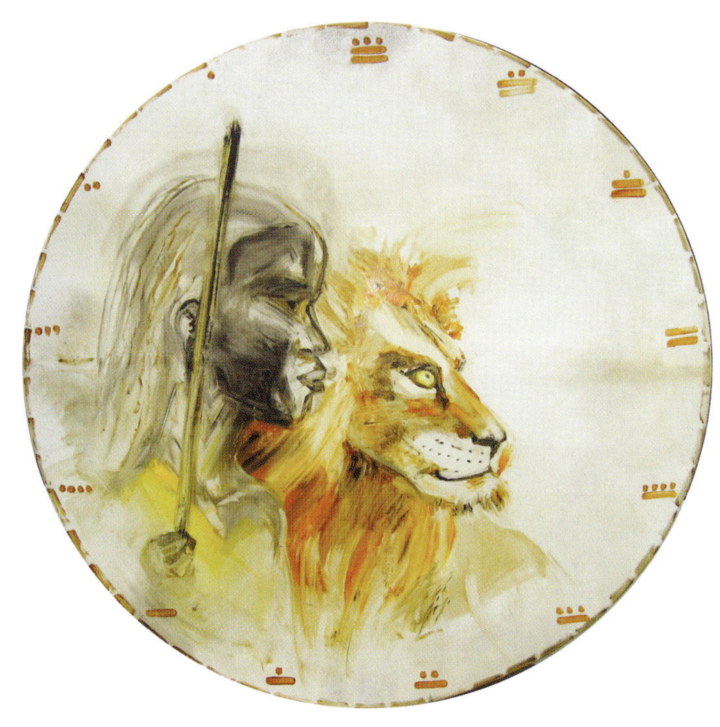

They Only Take What They Need

Oils on Goatskin Drum ø 95 cm

.... Western Ways

Most of us work at jobs,

far from the power to make a choice.

Jobs not worthy enough to invest most of our lifetime.

In the little time left,

we try to make up.. with the money earned,

having a hard time to find happiness.

FORGETTING THAT NATURE PROVIDES ALL WE EVER NEED,

FOR FREE

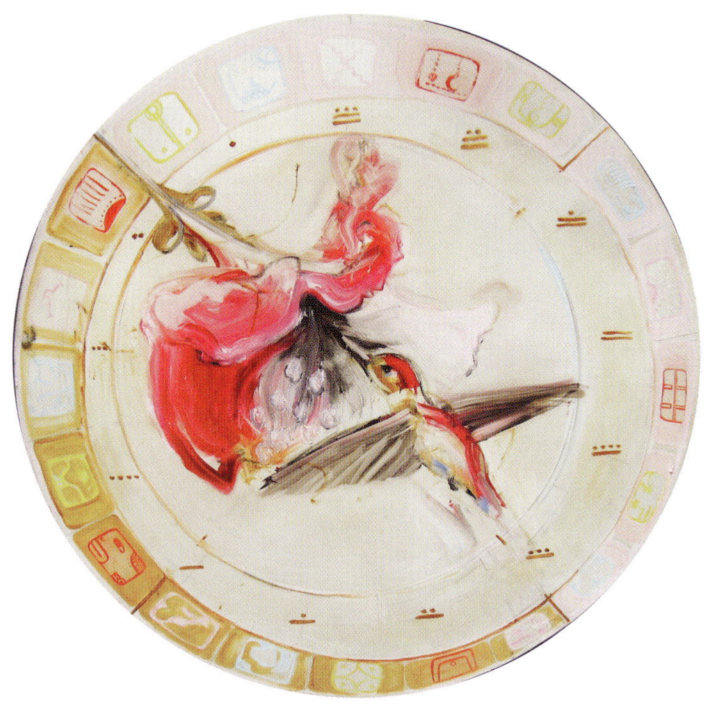

Divine Moments

Oils on Goatskin Drum ø 85 cm

.... Western Ways

We teach each other

to live life to the fullest.

..So we go for it all.

And when you are one of the luckiest,

you get your hands on all the MustHaves!

Surprisingly, the winner keeps on searching,

FOR 'THAT ONE THING' WHICH MAKES IT ALL WORTHWHILE.

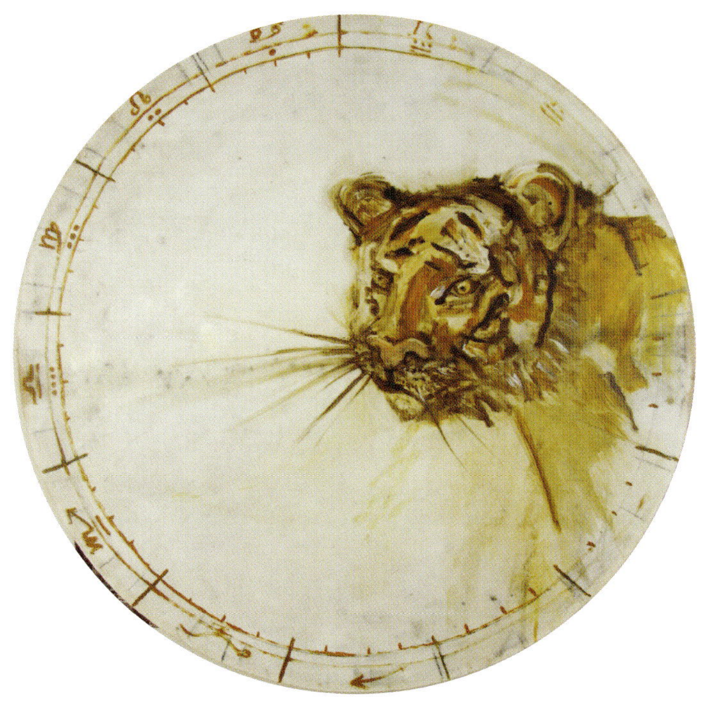

Time is Running Out
Oils on Goatskin Drum ø 75 cm

.... Western Ways

To live a fulfilling life today,

is increasingly difficult,

in this crazy world we created.

Too many mind-made-rules to follow.

Too many waiting queues to get ahead.

THE OUTSIDE WORLD SHOWS US CLEARLY,

WE ARE LOST IN A LABYRINTH.

— Power

We humans have been gifted with the power

to look after all the others on Earth:

the Stones, the Plants and the Animals.

We are free to have a taste of anything.

Selfishly, we only seek to fulfil our own wishes.

We misused the freedom.

We misunderstood the power.

Heaven on Earth
Oils on Goatskin Drum ø 50 cm

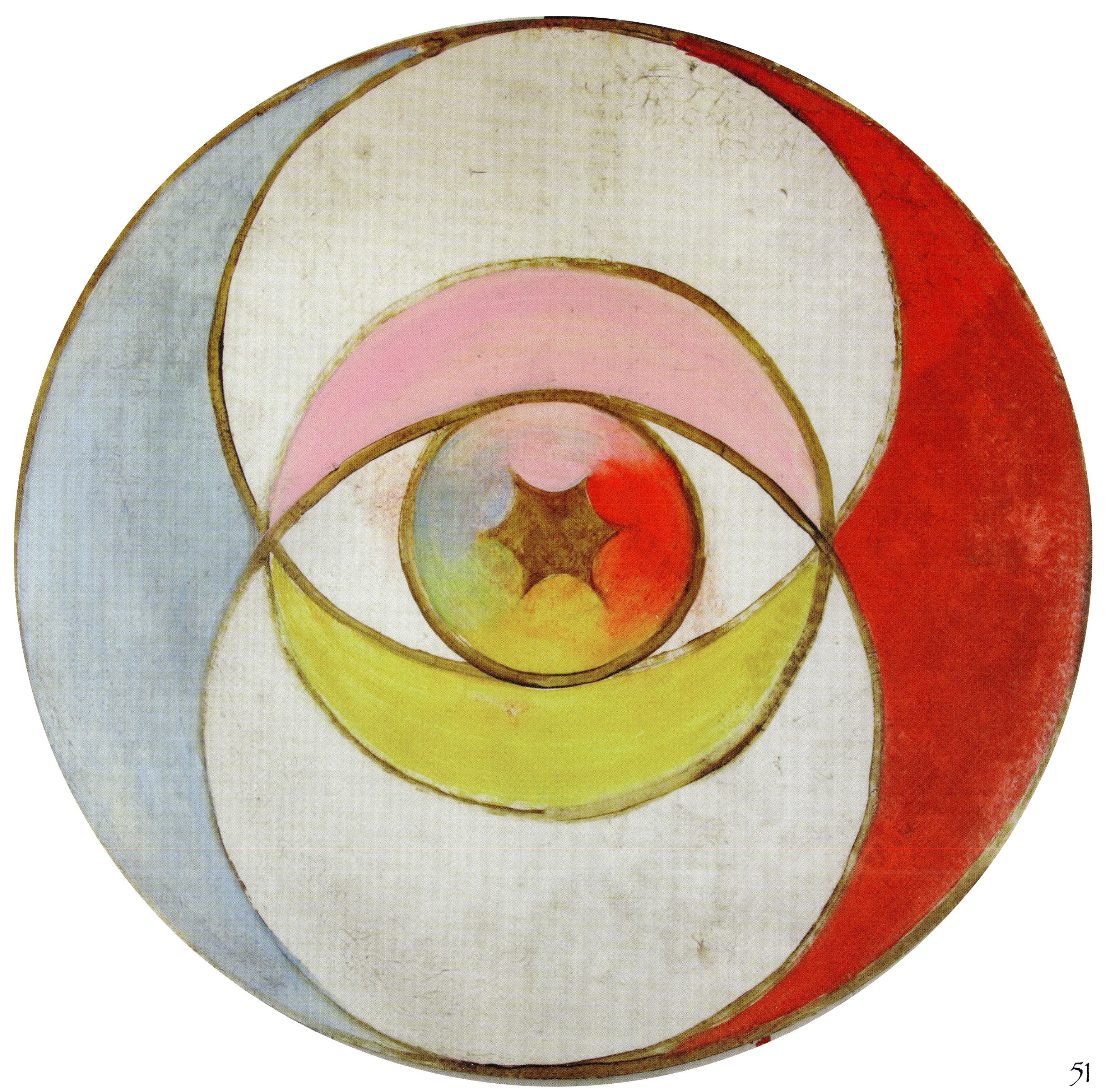

– Power

The freedom to have,

but not to hold on to.

The power to care,

but not to control.

Free just to enjoy who you are

and to appreciate the others' otherness.

That is real power set free.

Tree of Life

Oils on Goatskin Drum ø 50 cm

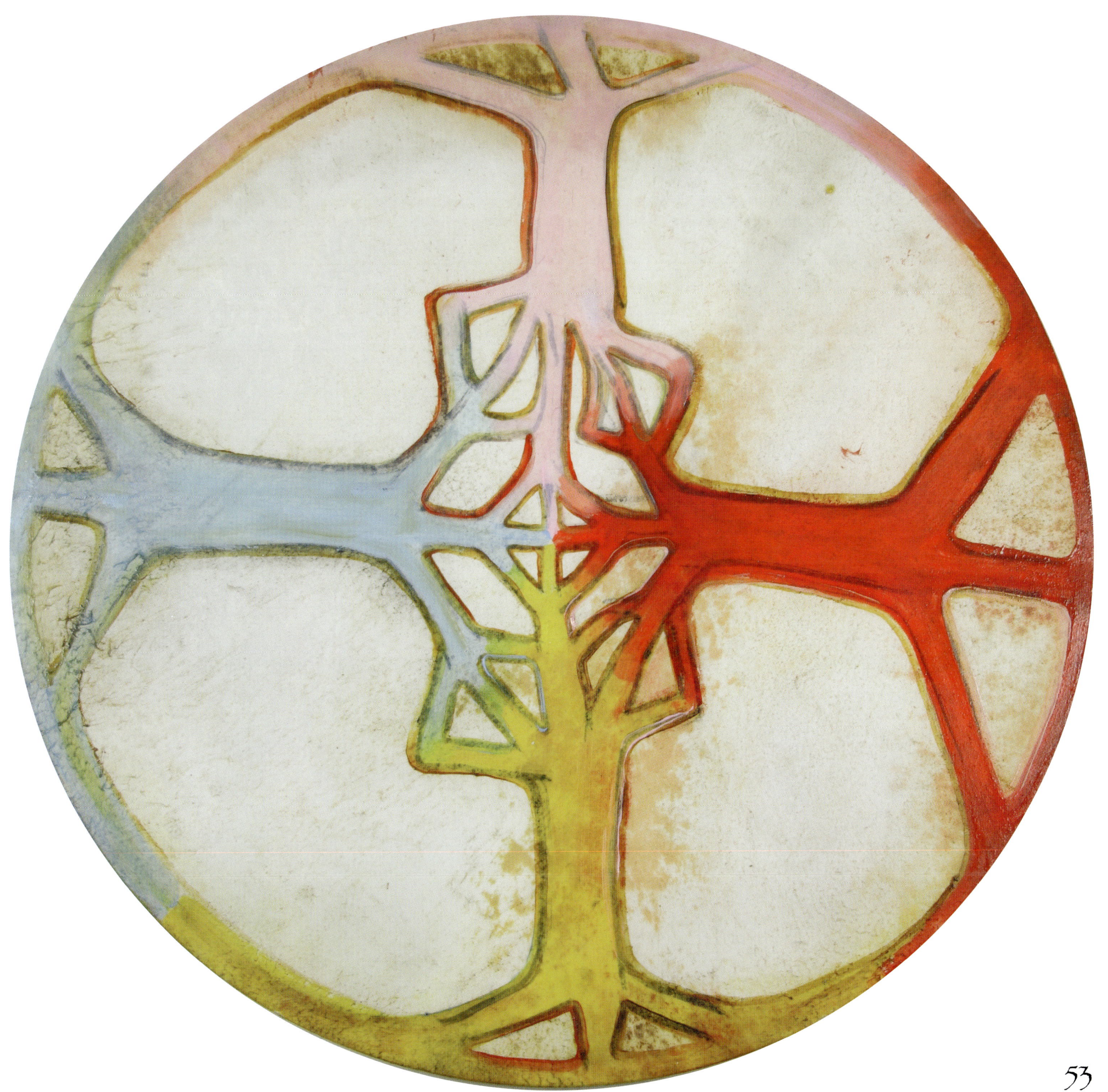

— Power

Among ourselves we naturally compete.

The best should be leader in their field,

as a leader you may show others the way.

YOU ARE HIGHLY GIFTED TO OCCUPY A MAJOR PLACE.

Others will make you feel important.

You have the power to do as you choose.

Time will tell how you used your talents

Why You of All

Oils on Goatskin Drum ø 50 cm

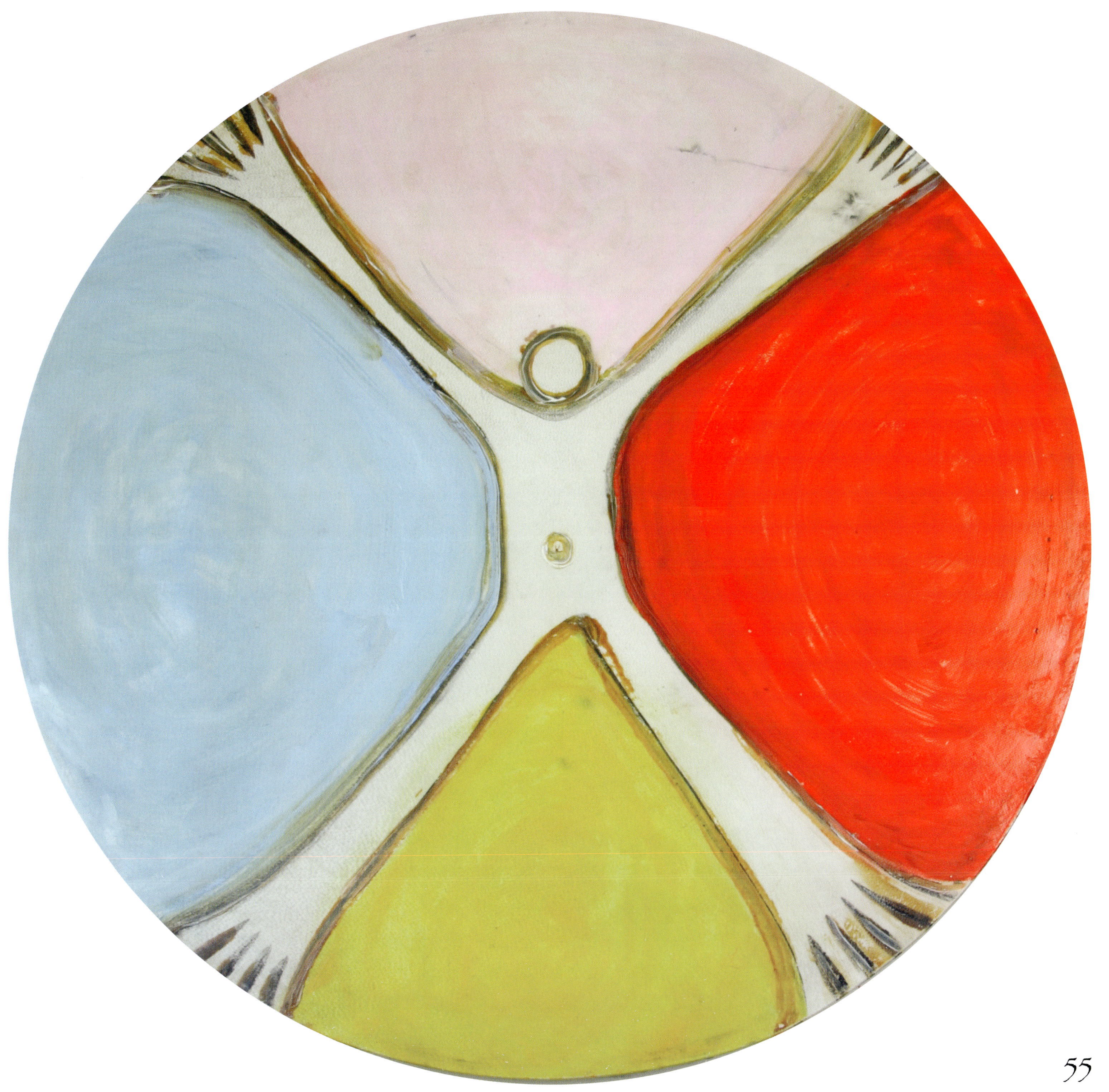

— Power

THE ART OF LIVING

IS KNOWING WHAT TO DO

AND WHEN TO LET IT BE…

It is giving full attention

to your real interests.

Most likely,

that is what you came here for, anyway.

Intention

Oils on Goatskin Drum ø 50 cm

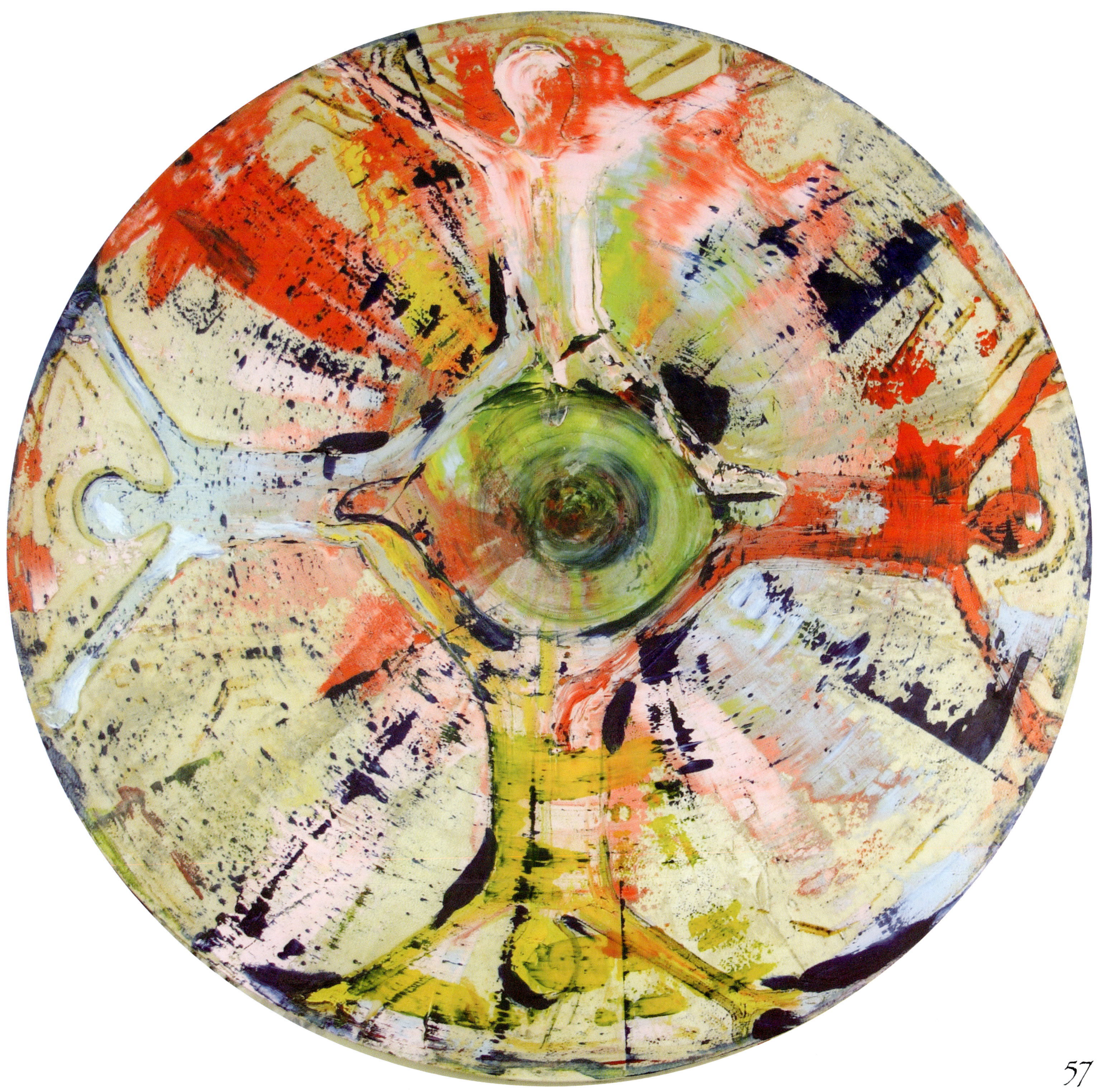

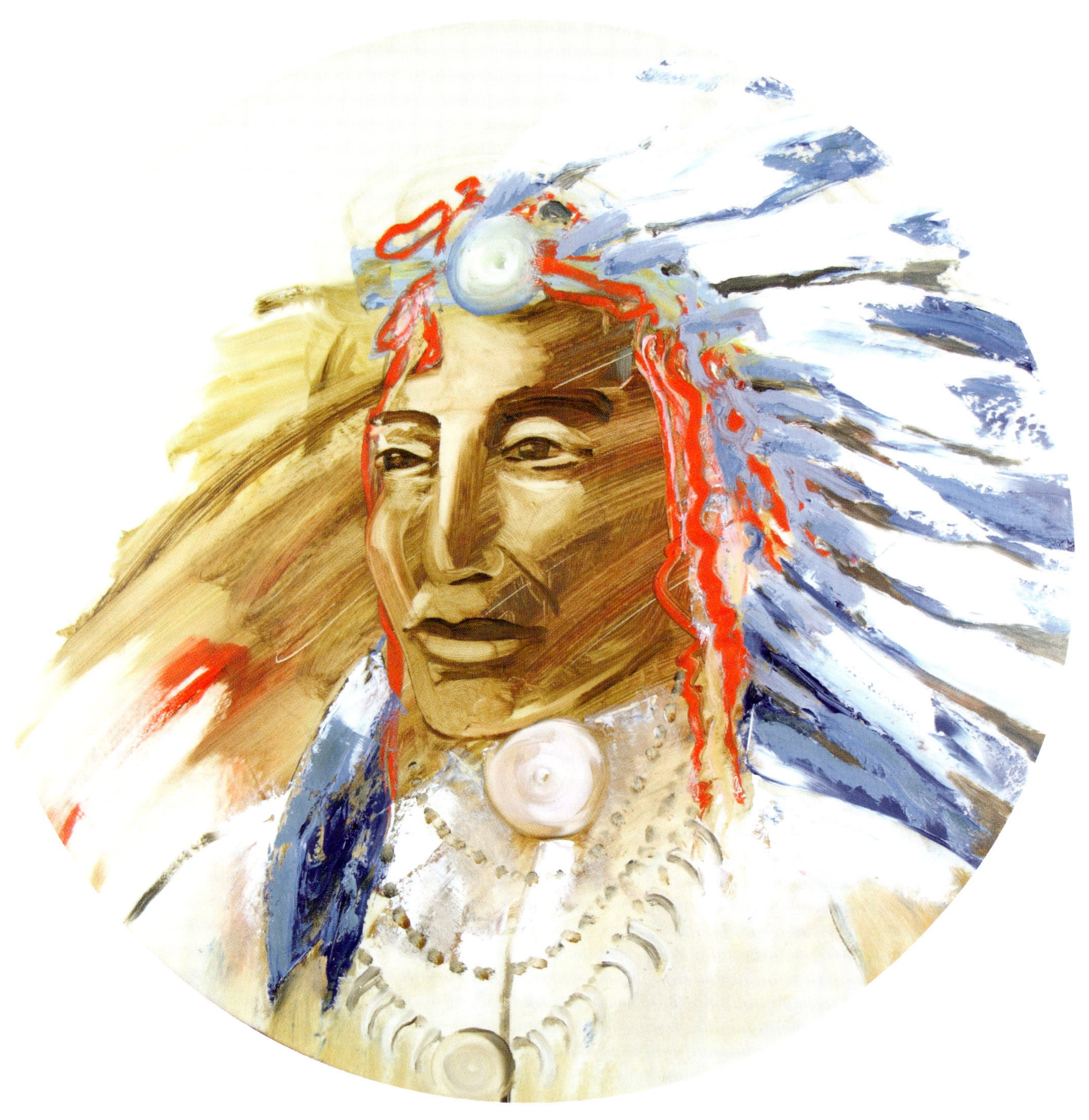

Chief
Oils on Goatskin Drum ø 70 cm

Animals & Plants

The Indians love the spirit of life

and respect it equally, in all various forms.

THEY SPEAK FROM SOUL TO SOUL.

They know Trees can talk,

flowers can sing

and that Mountains live long enough,

to know almost everything.

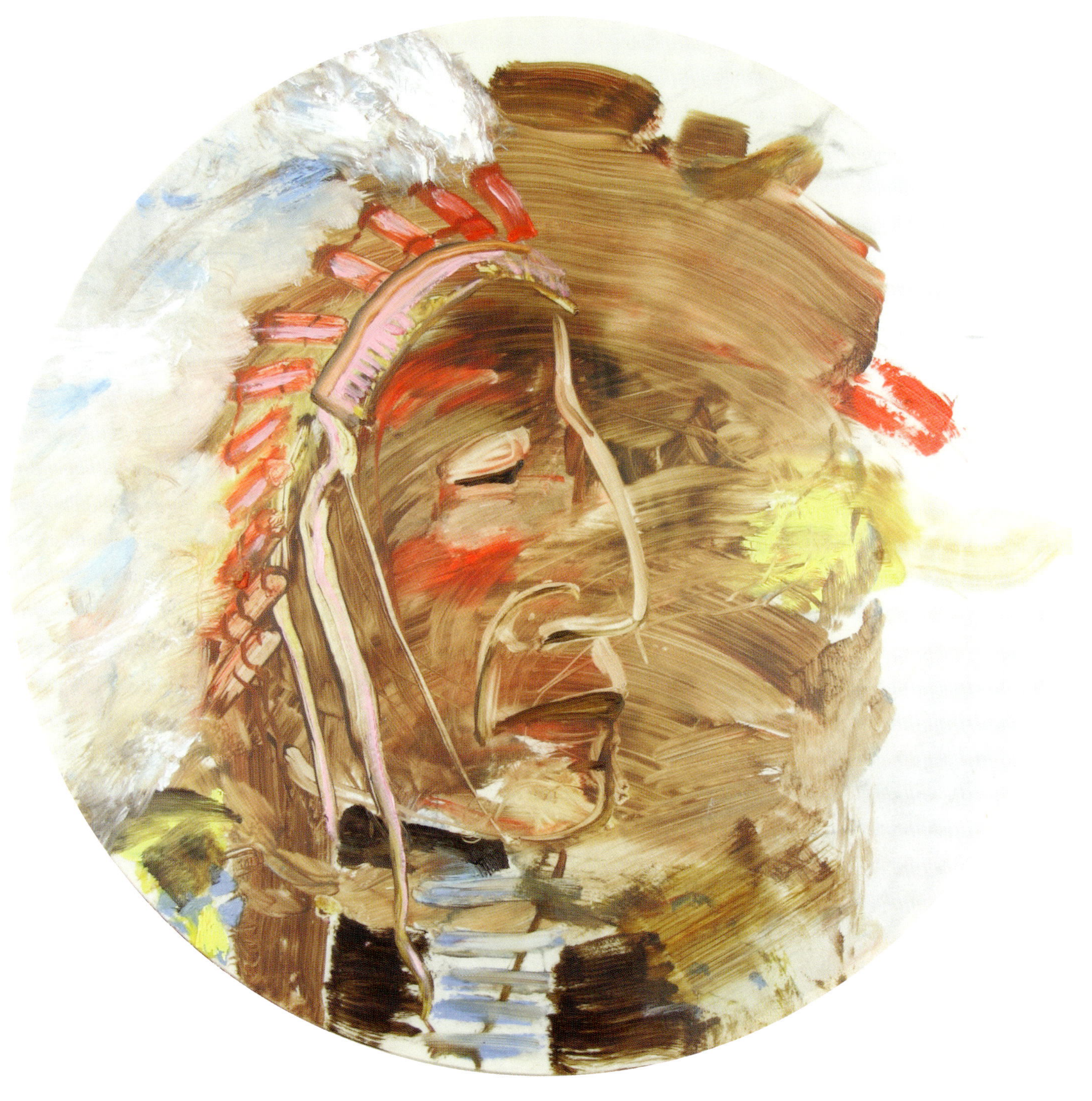

Medicine Man

Oils on Goatskin Drum ø 70 cm

∸ Animals & Plants

The Wind whispers messages from afar.

The Rain flushes the superfluous away.

The Clouds change your thoughts.

The Sun warms your soul.

The Moon gives you insight.

AND THE SKY SHOWS WHICH PART YOU PLAY,

TO MAKE UP THE WHOLE!

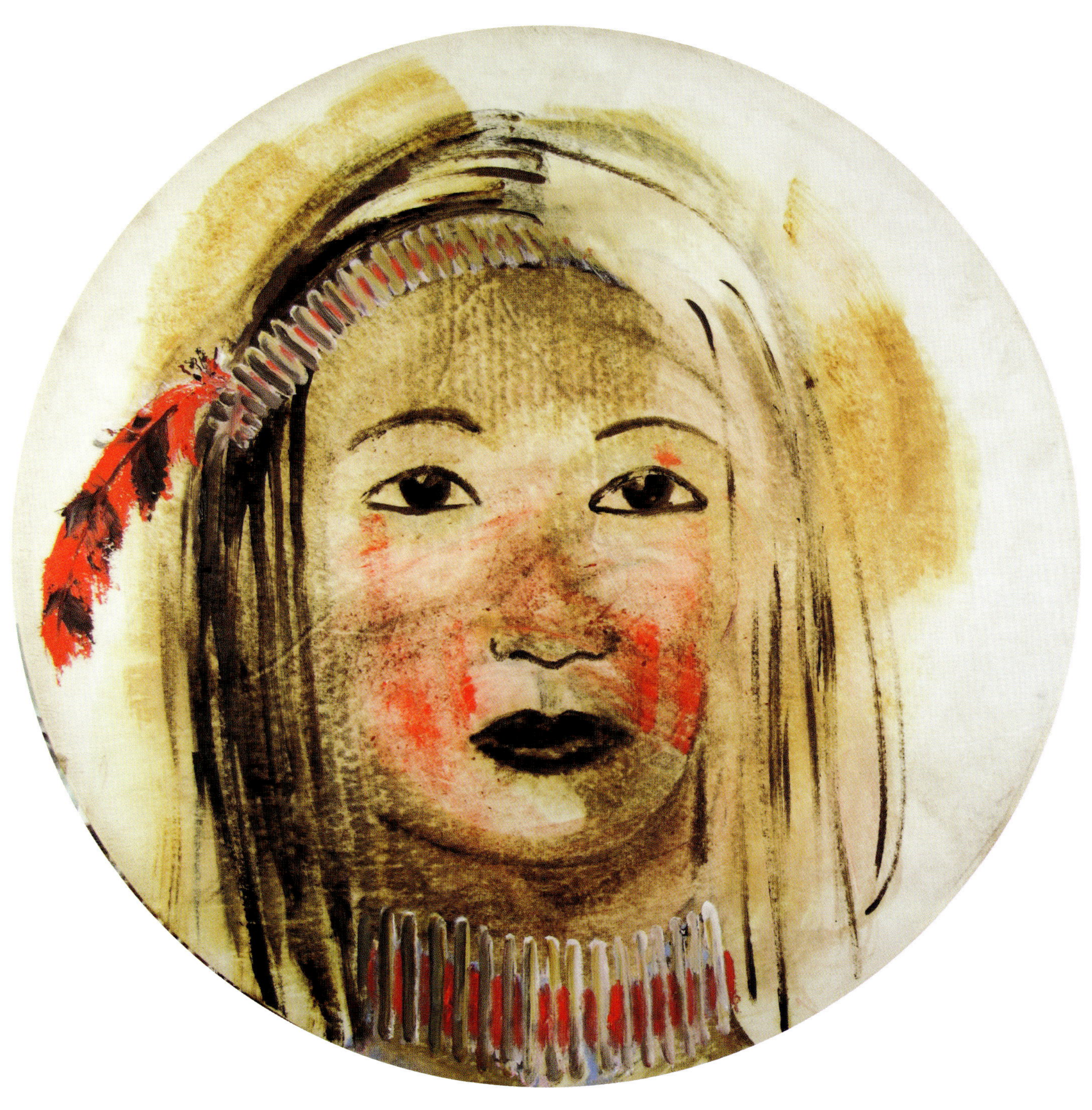

A View Ahead

Oils on Goatskin Drum ø 70 cm

∸ Animals & Plants

The Eagle has the overview.

The Mouse dwells on details.

The Dragonfly colors long forgotten worlds.

The Lizard believes in your fantasies.

The Beaver shows how to build them.

And the Horse gives you the power,

TO MAKE YOUR DREAM COME TRUE!

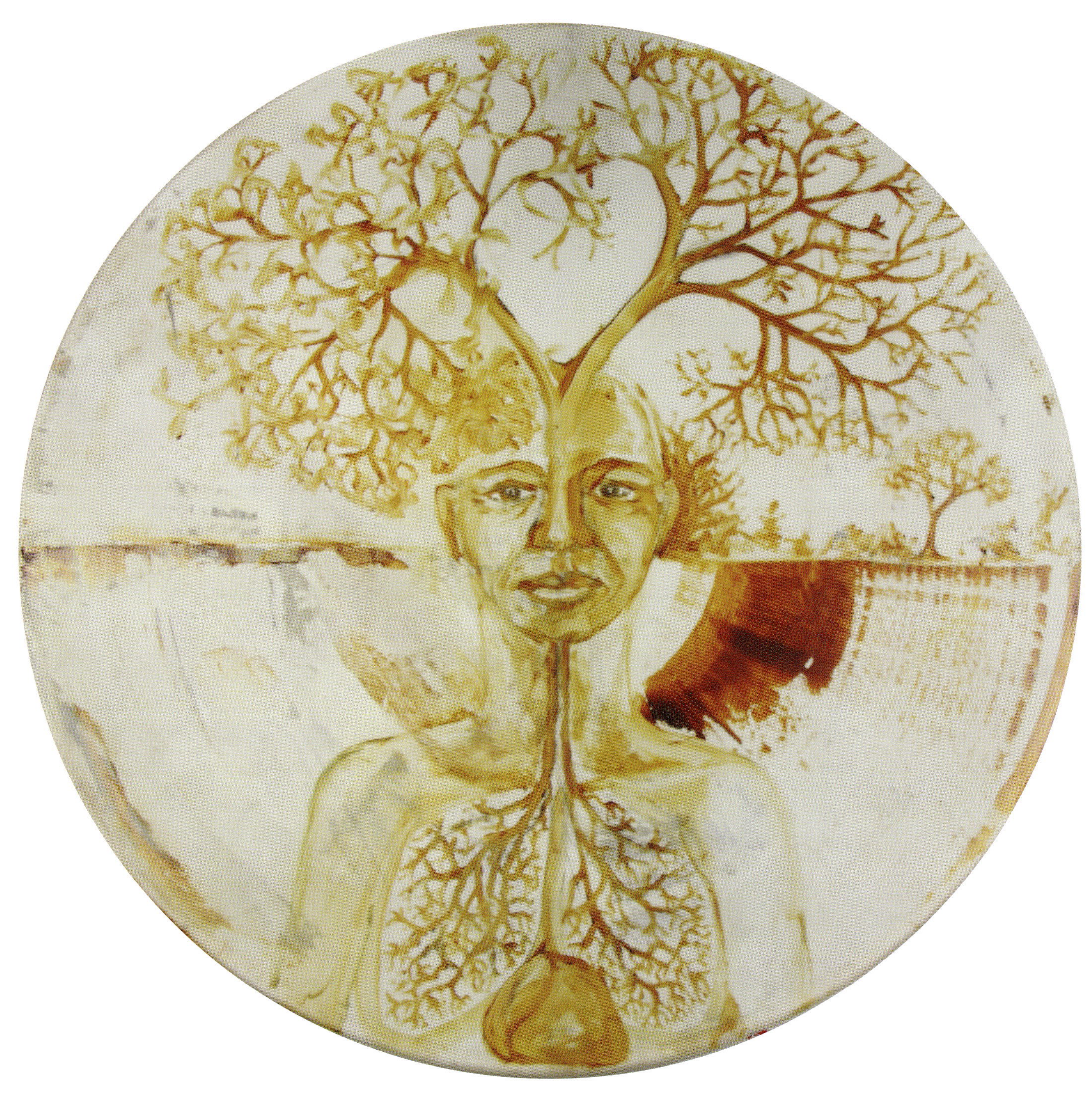

Breath-taking Trees

Oils on Goatskin Drum ø 50 cm

∴ Animals & Plants

The Oak knows all about love matters

Those strong feelings straight from the heart,

which make the blood run faster.

The Beech teaches about self confidence

Massive trunk, hardly any roots, but proud to be frank and loud

The Birch gives hope,

it is the first to reappear after a devastating fire.

Place

But we Westeners walked over the Indians

Killed almost all, not so long ago

And now it seems the Indian Way of Life,

was more advanced than ours today

How strange with all those wonderful inventions,

we have in our modern lives today;

We even walked on the Moon and touched Mars.

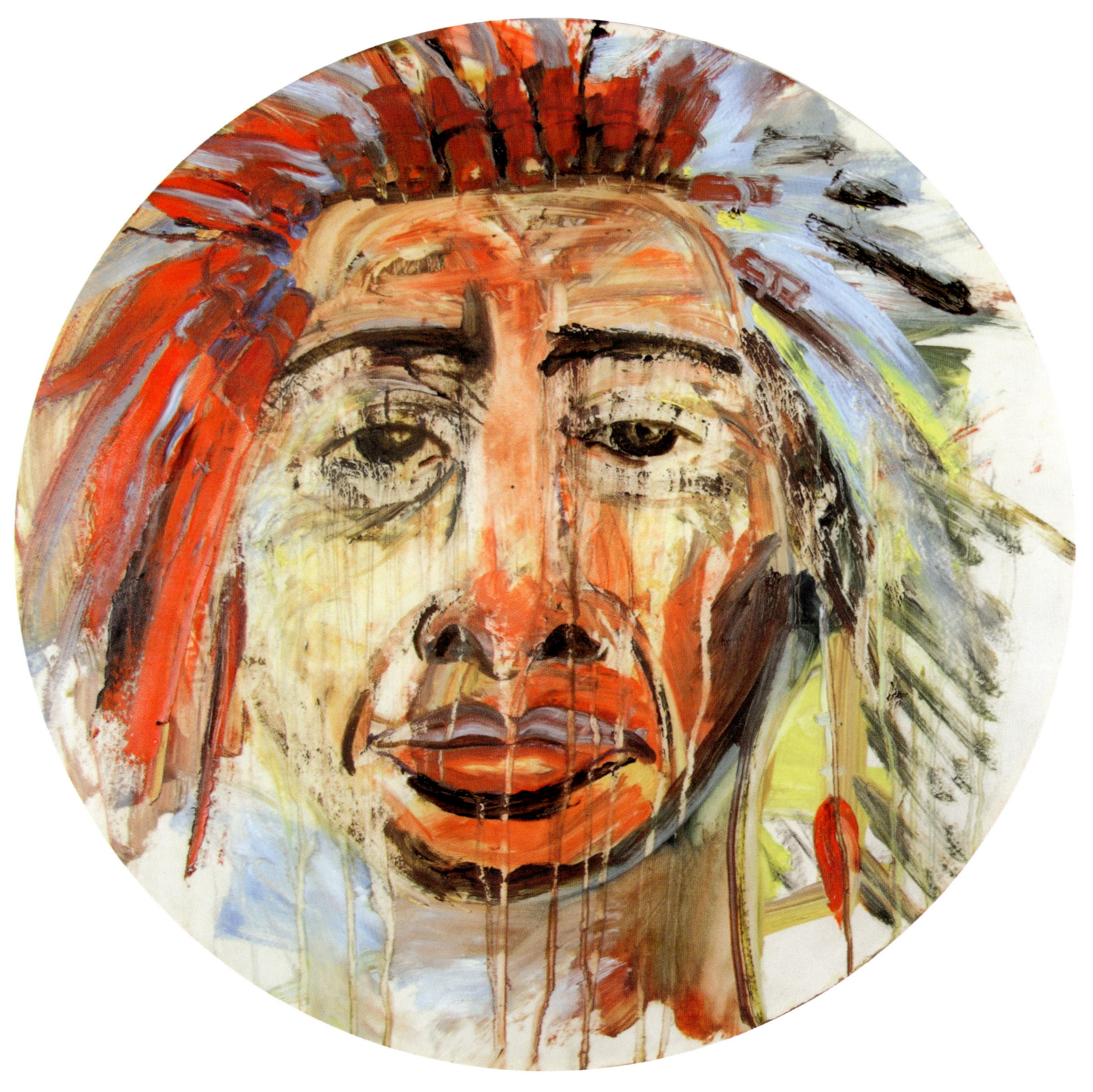

Devastation
Oils on Goaytskin Drum ø 70 cm

⚍ Place

Still, the Earth is the only place

where we can live in space

AFTER 200,000 YEARS OF BEING HUMAN,

WE STILL DON'T SEEM TO KNOW HOW TO CARE

FOR HER, THE EARTH.

Like any bad relationship,

it is sickening both beings.

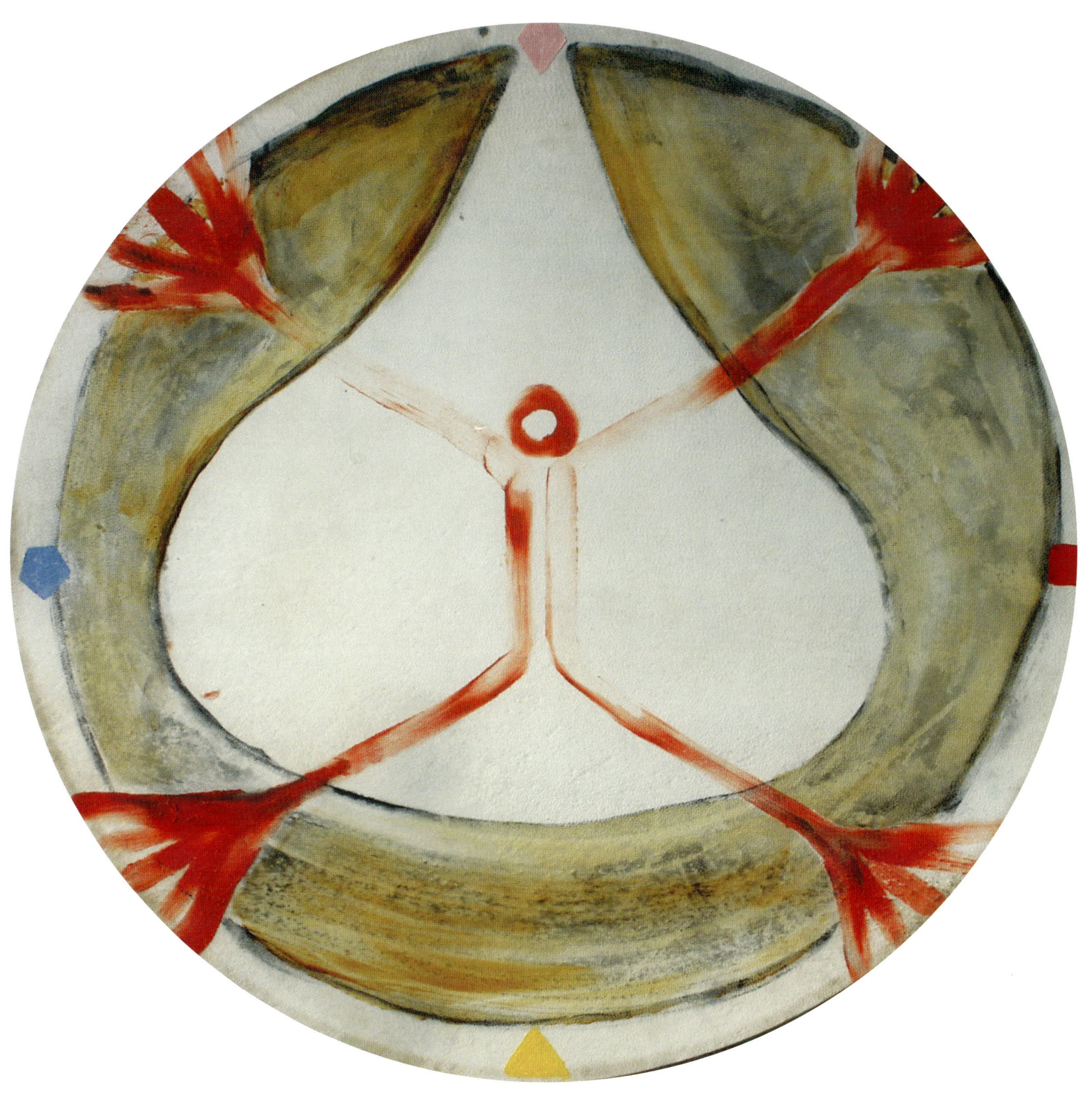

Mankind
Oils on Goatskin Drum ø 50 cm

⩨ Place

We humans lost sight of proportions

by overestimating ourselves.

From a spacious viewpoint,

EARTH IS THE BIG IMPORTANT ONE.

And we are just a bunch of lucky fellows,

not valuing her enough for the place she takes.

You ask yourself: Why on Earth..?

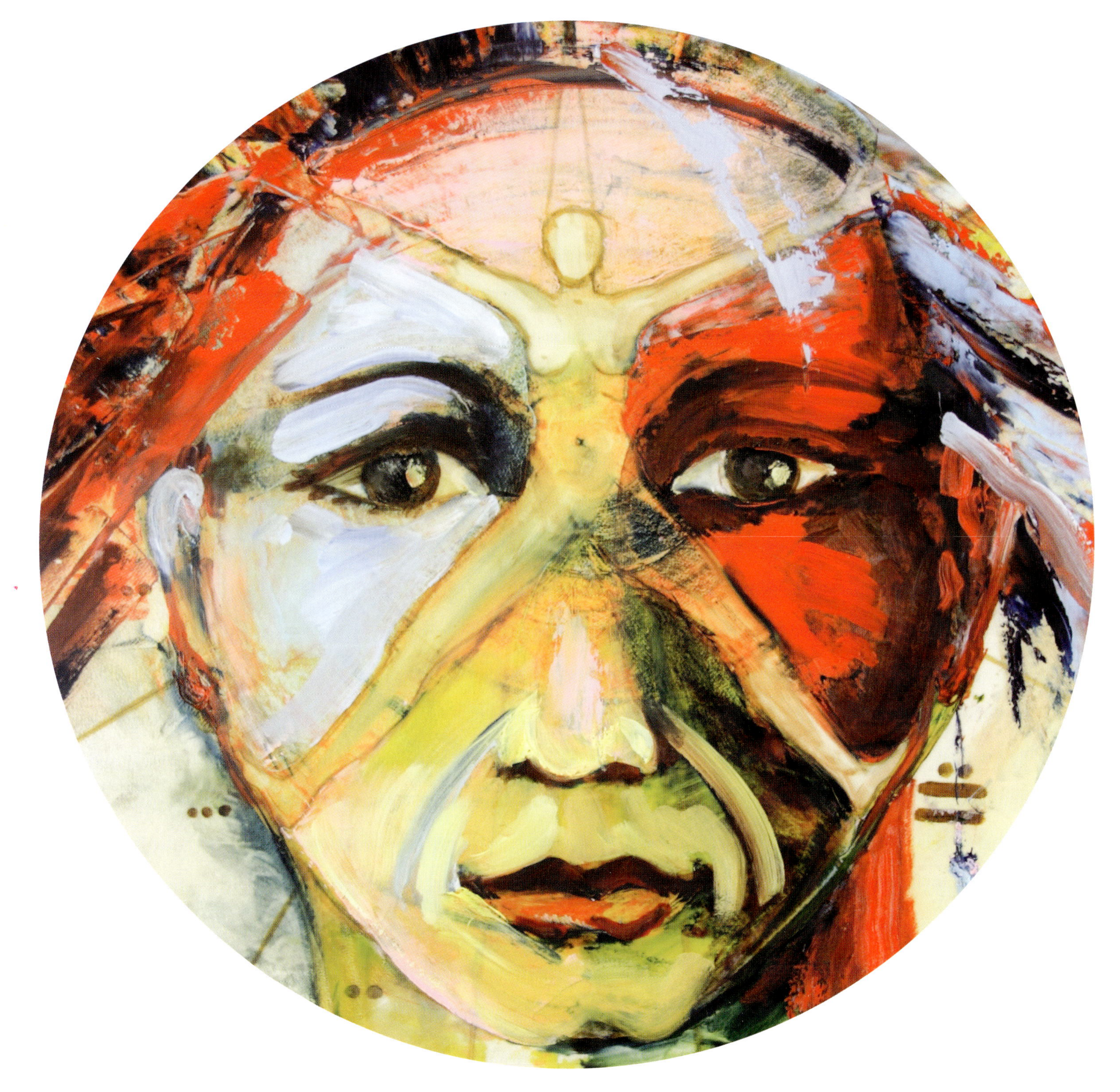

Natural Order
Oils on Goatskin Drum ø 50 cm

·· Place

Considering the Sun to be most important in the sky,

reflects our dominant male society and its focus on science.

Recognising the Moon's influence

will increase our intuitive senses,

like: ~ *what dreams at night try to tell you* ~

~ *gut feelings, how to act from the heart* ~

~ *and to listen to your intuition* ~

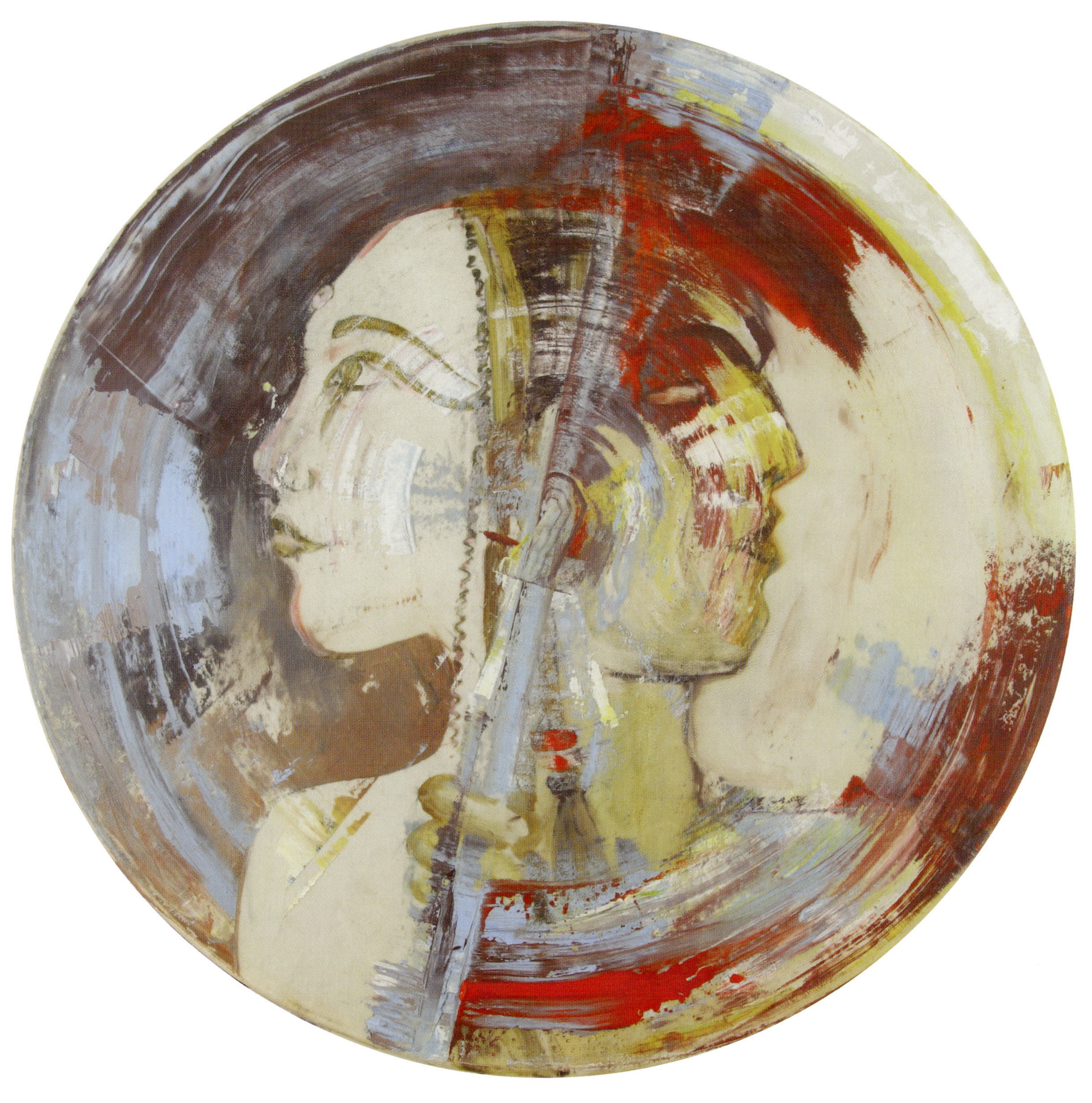

Perfect Opposites
Oils on Goatskin Drum ø 50 cm

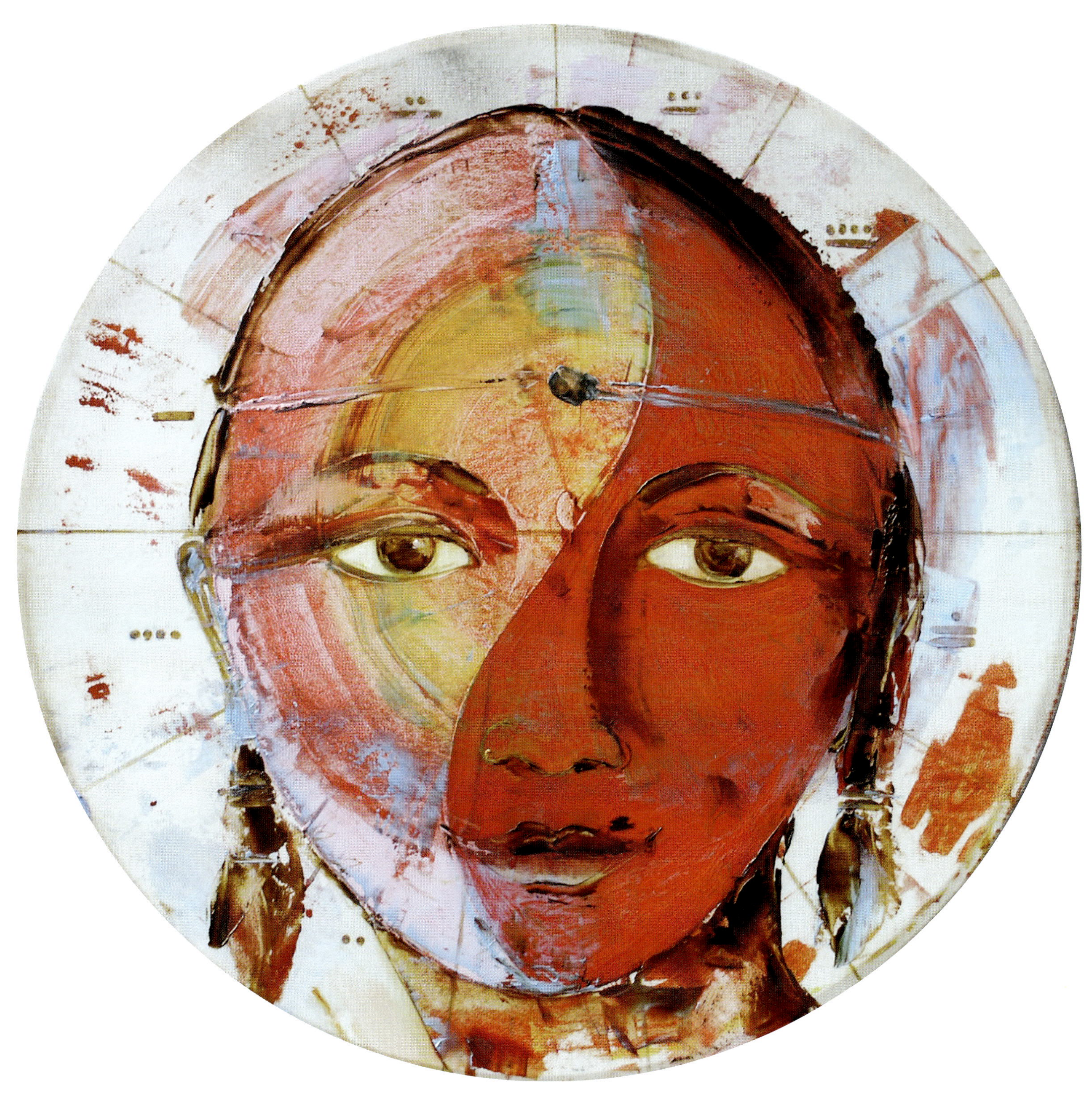

Full Moon

Oils on Goatskin Drum ø 50 cm

··· The Moon

She turns heads,
    the Moon,
with her different faces,
    her intuitive character
        and hidden side.
SHE TURNS AROUND YOU,
    WAITING FOR YOUR ATTENTION.

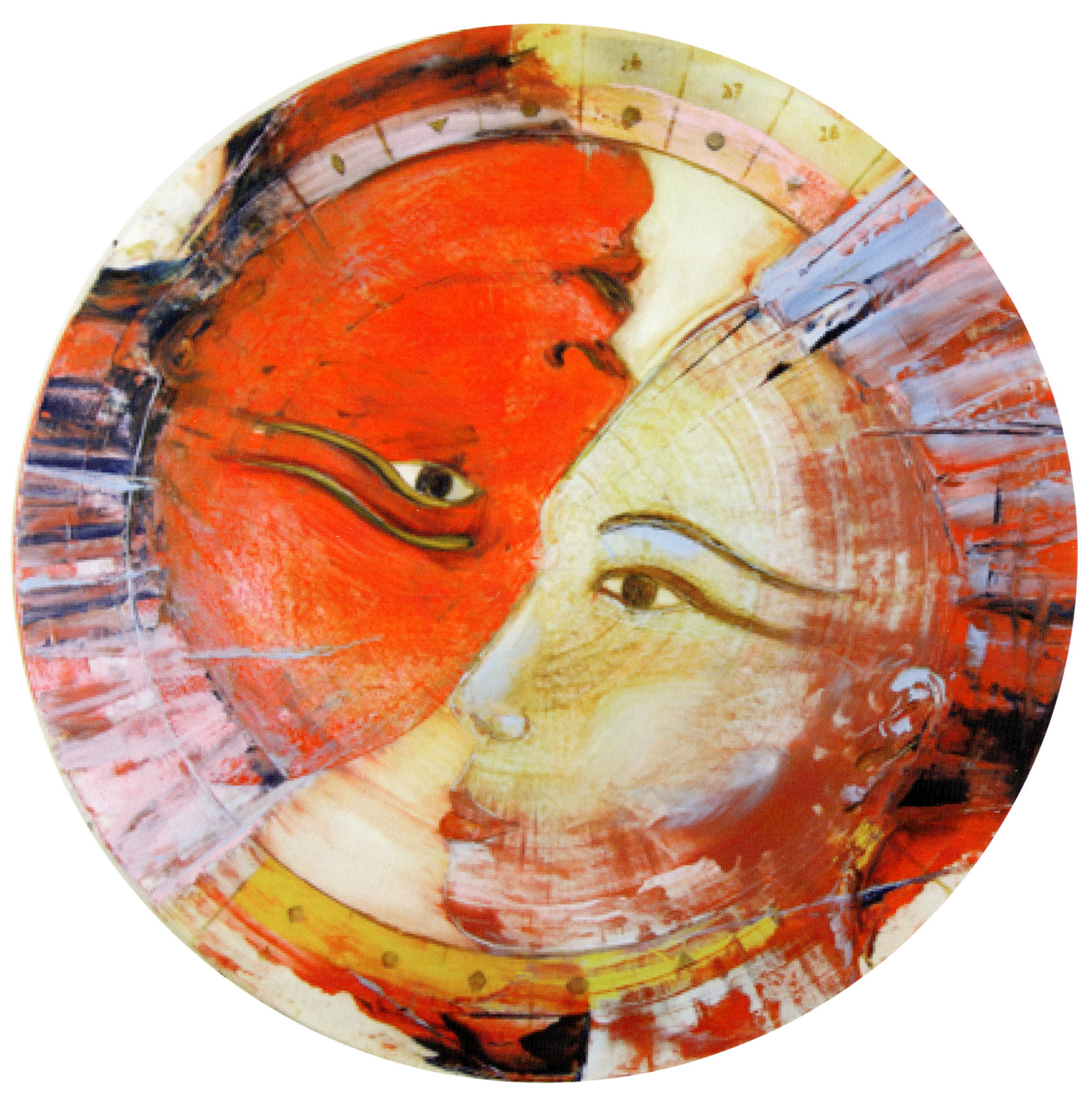

Moon Waves

Oils on Goatskin Drum ø 50 cm

⠶ The Moon

THE TURN OF THE TIDES

ARE CLEAR FOR EVERYONE TO SEE

To help you find the right time,

to go with the flow,

like a wave.

It comes and goes,

time and again.

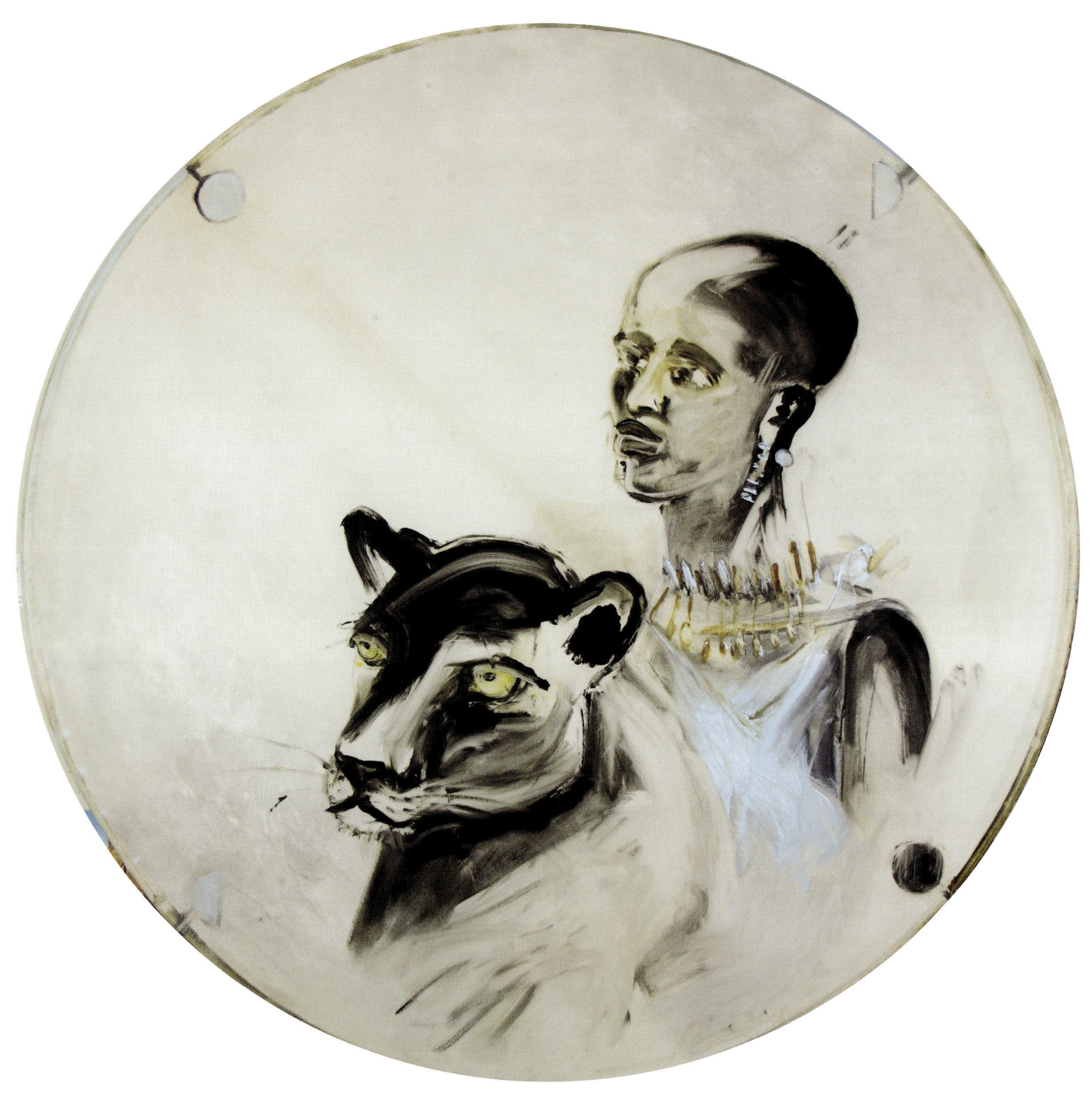

A New Moon

Oils on Goatskin Drum ø 95 cm

... The Moon

THE RHYTHM OF LIFE

MOVES LIKE THE MOON

It grows, it builds up,

it becomes full Moon,

then the tide turns,

it shrinks, it lets loose

and reappears into a New Moon.

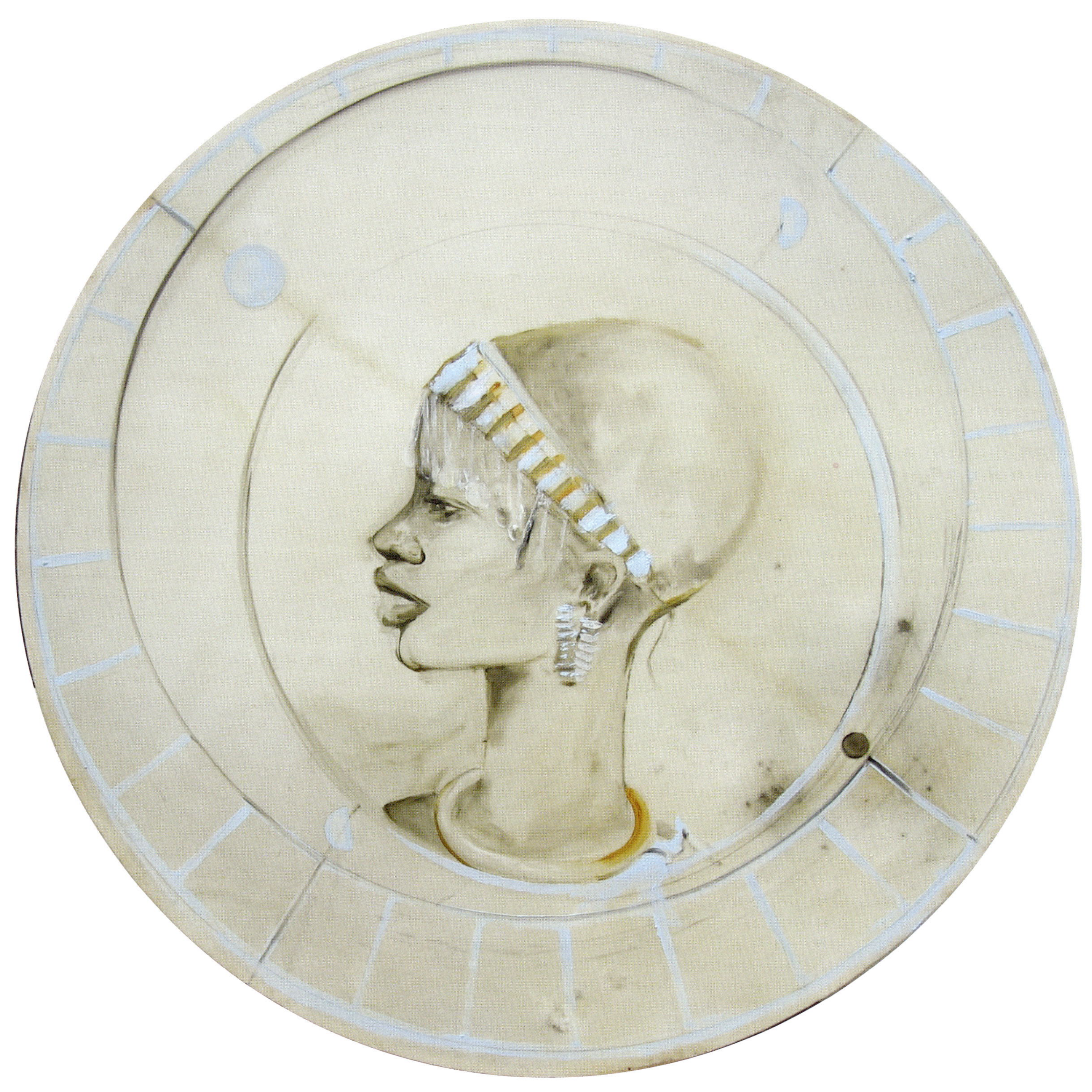

Influence of the Moon

Oils on Goatskin Drum ø 85 cm

∴ The Moon

The Moon indicates

the right time to take action,

to soil your ground,

to plant your seeds,

to water your plants,

to harvest your fruits,

and when to enjoy it to the fullest.

•••• The Sun

Sun and Moon

show up in the sky

same in size

to remind us

of their equal importance

to all existing life

on planet Earth

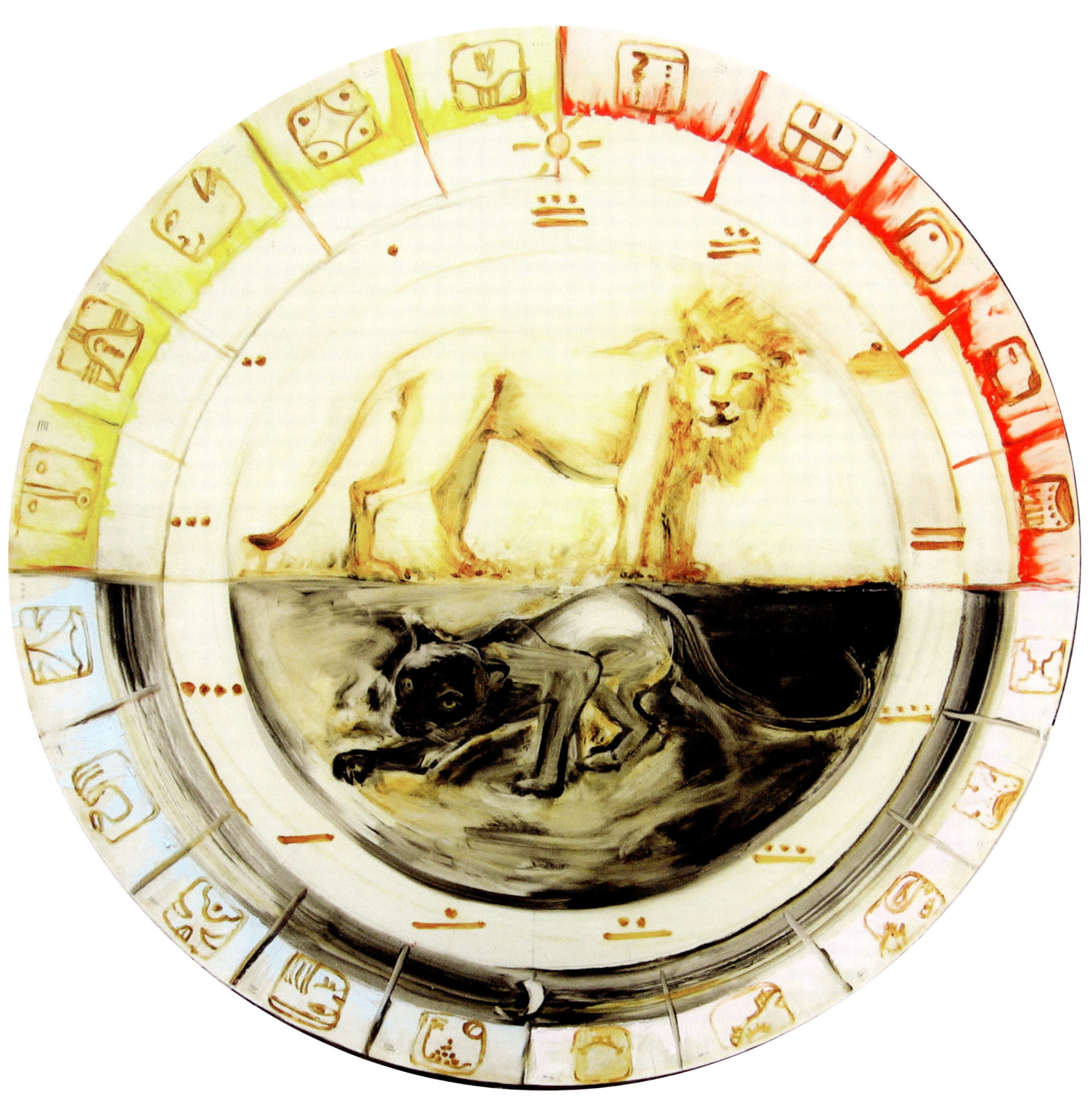

Day & Night
Oils on Goatskin Drum ø 85 cm

•••• The Sun

Sun and Moon

are matched for life,

bonded by love.

THEY RULE DAY AND NIGHT

DOMINATE SUMMER OR WINTER.

A pair beyond the clouds,

chasing each other around the Earth.

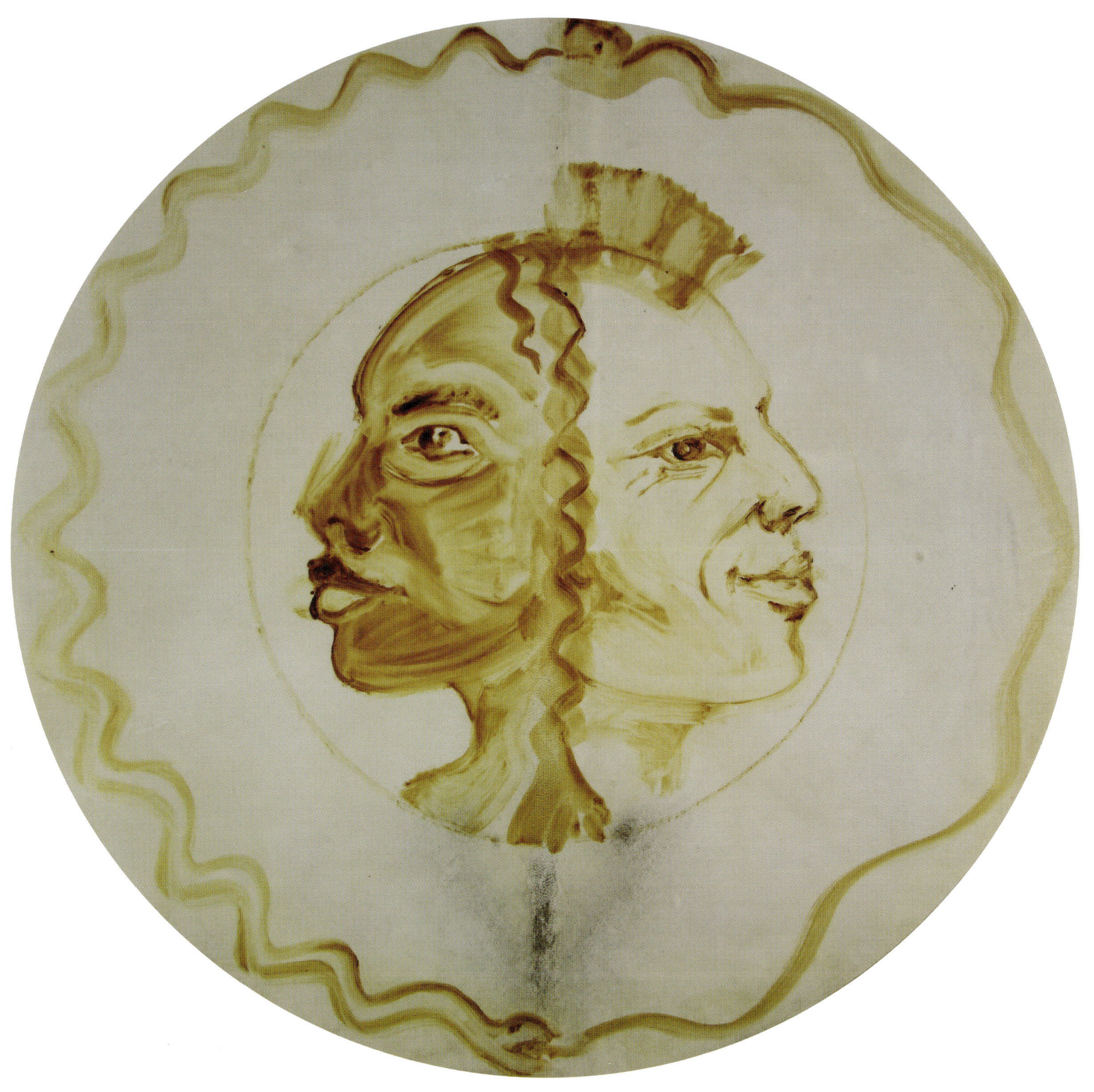

He & She

Oils on Goatskin Drum ø 75 cm

···· The Sun

Sun and Moon, love at first sight.

He shines, clear about his intentions.

She giggles, teases, sending mixed messages to size him up.

She becomes his prime focus, he jokingly suggests:

"Forget about shadows, lets make love in bright daylight."

She giggles, teases some more and slowly but surely she seduces him.

ECSTATIC LOVESPOTTING PUTS SUN & MOON IN HEAVEN.

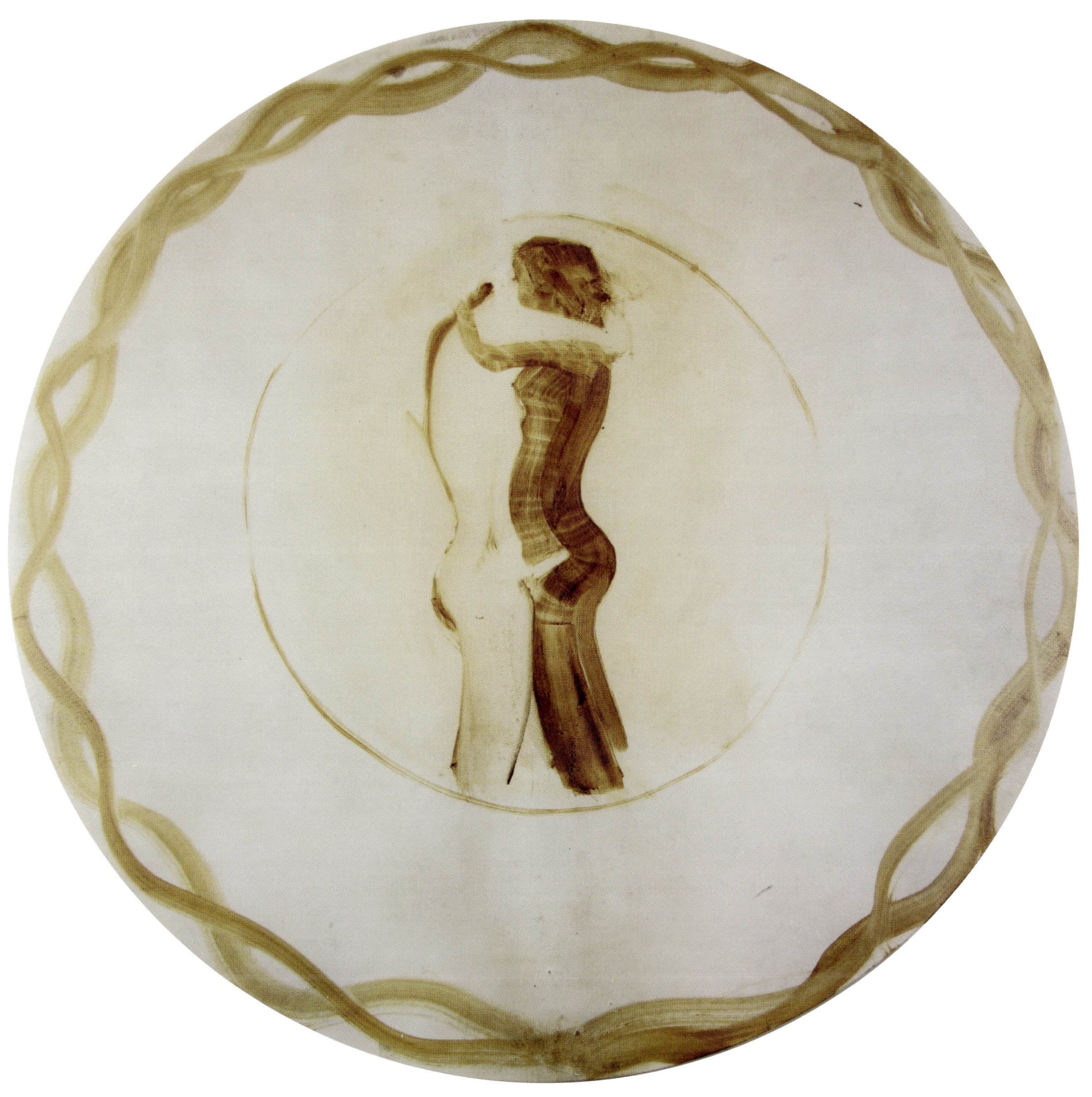

Love-spotting
Oils on Goatskin Drum ø 75 cm

•••• The Sun

THE SUN OUR STAR,

source of heat,

source of light,

source of life,

source of happiness,

source of knowledge,

beamed up directly from the Source.

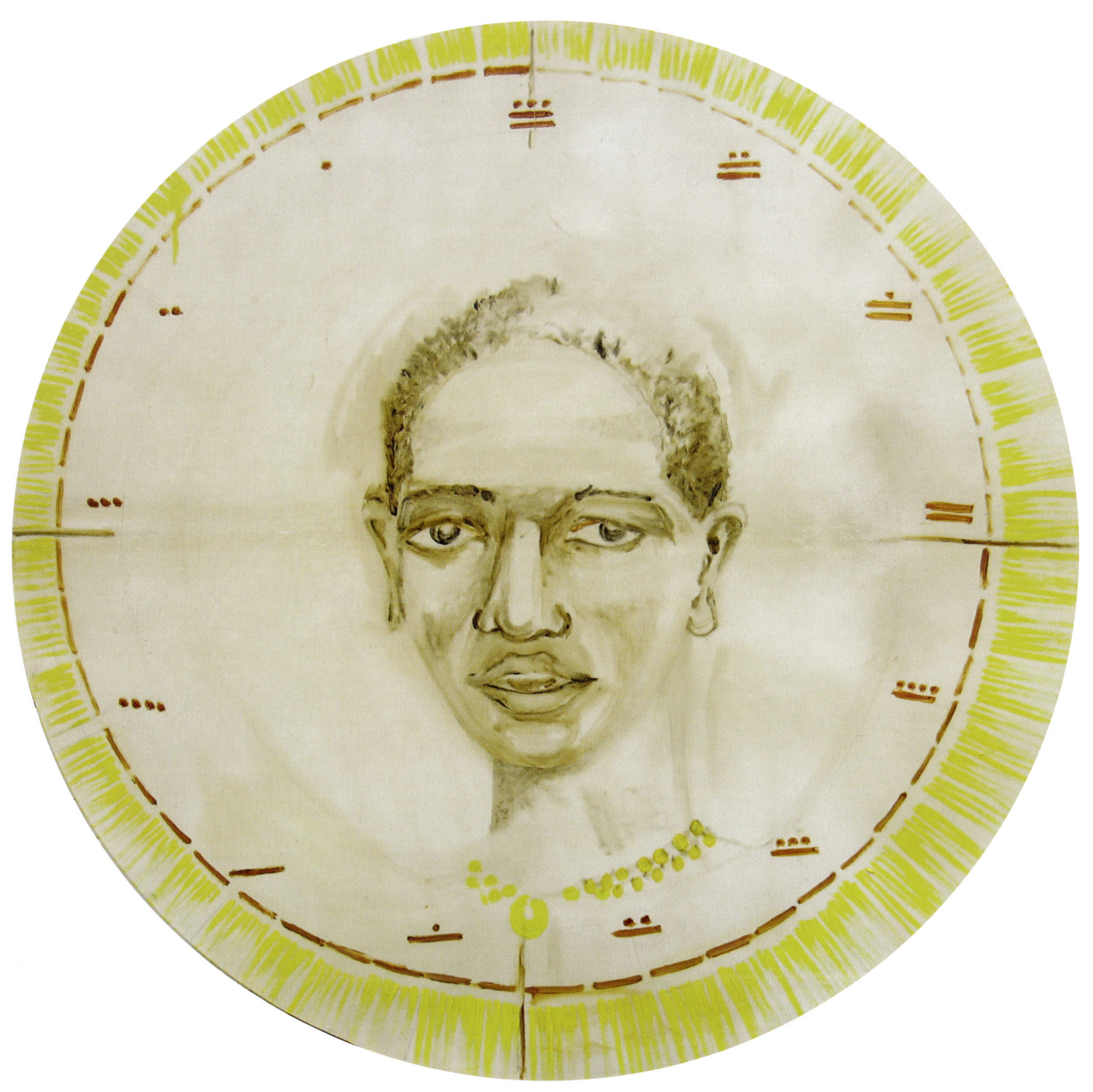

Africa Source of Origin
Oils on Goatskin Drum ø 85 cm

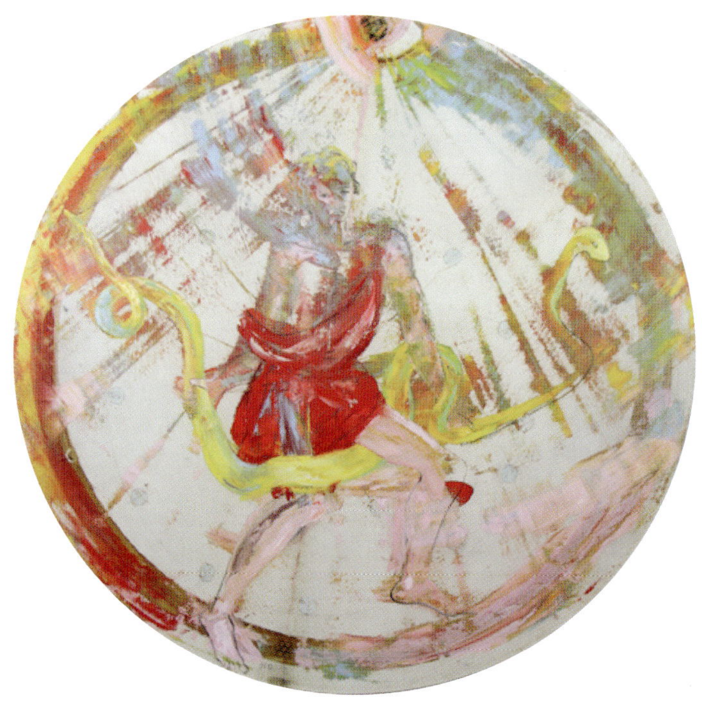

Ophiuchus
Oils on Goatskin Drum ø 85 cm

═ Maya

Now is the time to look further

than our solar system.

To find the centre of the galaxy,

Sagittarius points towards it,

Scorpio is closest by

and Ophiuchus is bringing it to us:

THE WAVE~PATTERN OF A HEALING SERPENT.

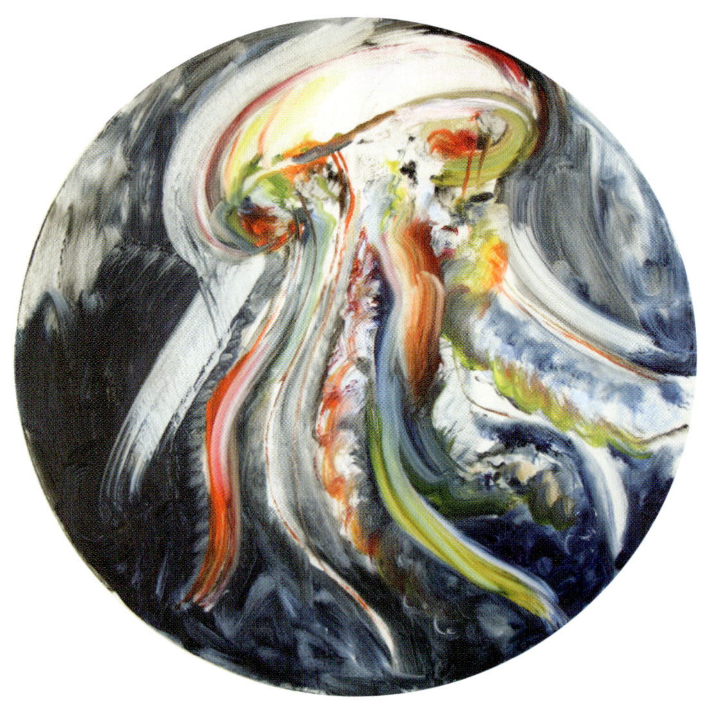

Universe

Oils on Goatskin Drum ø 60 cm

= Maya

The ancient Maya described

the Milky Way as a jellyfish of light,

having orgasms that make eternal waves

and pulsations in its field of attractions.

At its central axis: a wormhole, a black hole,

THE HUNAB KU:

THE ONE SOURCE & DESTINATION OF ALL

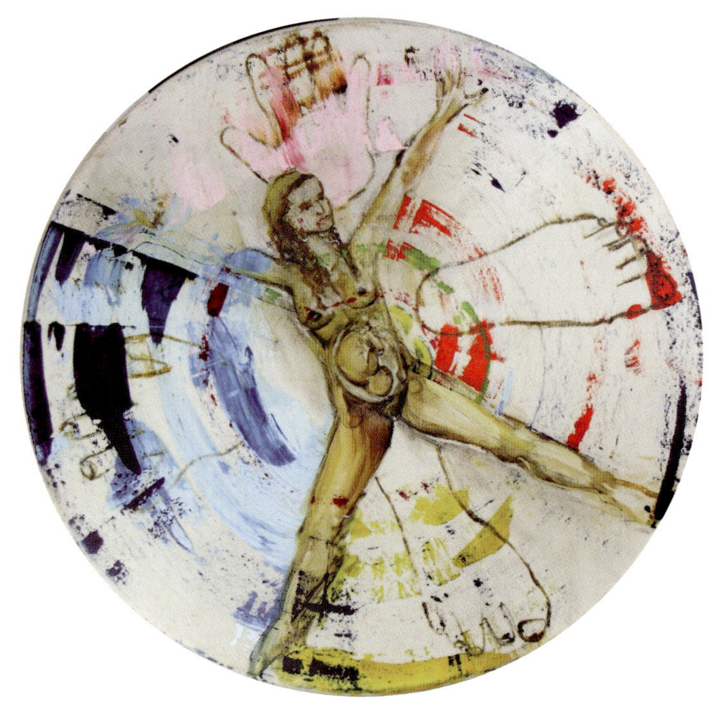

Tzolkin Calendar
Oils on Goatskin Drums ø 50 cm

= Maya

The ancient Mayas had 22 calendars.

They knew our Sun orbits the Pleiades in 26,000 years.

They knew Sirius, the Sun's twin, move together in space, like our DNA helix.

They knew the exact time Pluto's orbit takes.

They had an idea where we would be in the milkyway today, 2,500 years on.

THEY COUNTED THE DAYS 2,500 YEARS AHEAD, THEN STOPPED ABRUPTLY AT 2012.

..So what?.. You may think.

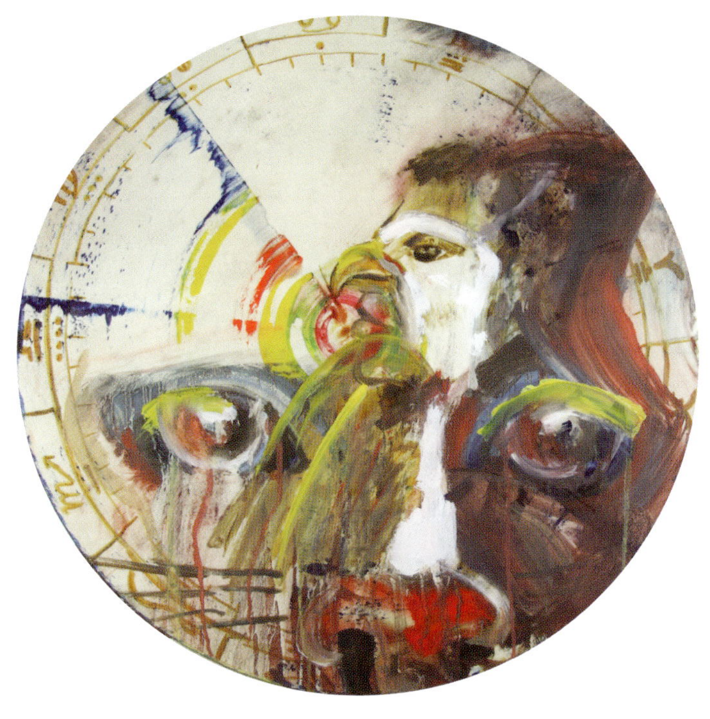

'MEN' Skywalker
Oils on Goatskin Drum ø 50 cm

= Maya

The Mayas build their Temples because of it

The Inca's digged their Nasca-lines to imprint it

The Egyptians placed their Pyramids to locate it

The Kelts constructed Stonehenge to time it

The Khmers designed Angkor Wat to point towards it

ALL FOCUSSING ON THE BIGGER PICTURE OF LIFE ON EARTH

And the wonders beyond....

Nexus

You cannot turn your head away.

It takes your breath away and evaporates every urge to create evermore.

Those masterpieces made overnight in the crop fields outside.

PERFECT PATTERNS SINGLE OUT THE CONNECTION WITH OTHER LIFE AROUND.

Just look at them: the patterns, the perfection, the size, the speed,

the message, the knowledge, the superb craftsmanship,

VIEWING ALL RELIGIONS AS EQUAL IMPORTANT TO FIND THE KEY.

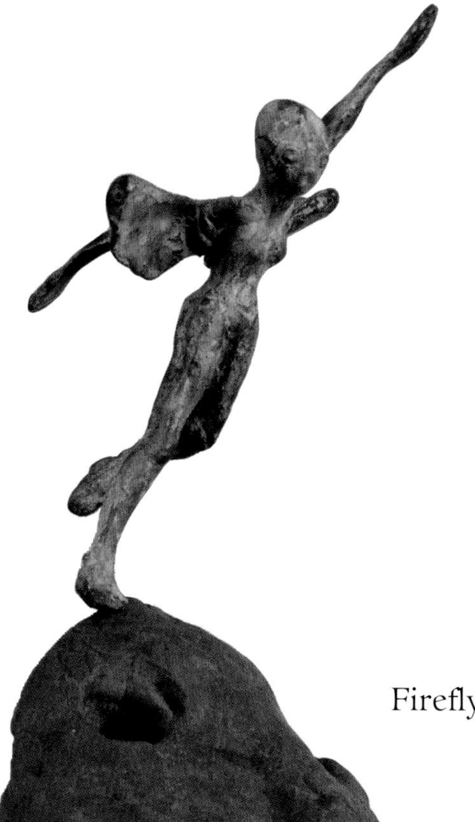

Firefly, bronze on boulder

Other Intelligence
Oils on Goatskin Drum ø 50 cm

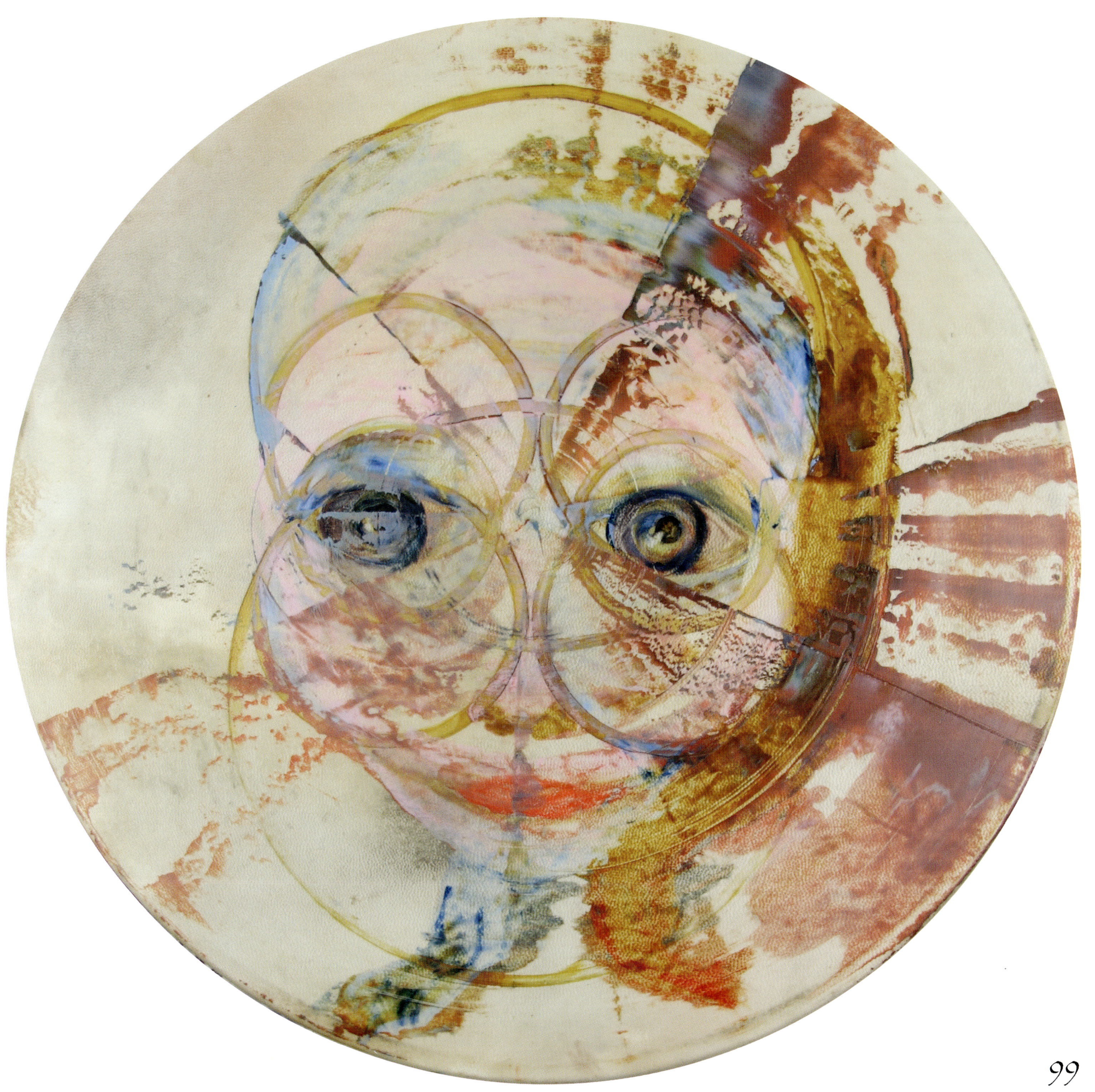

≐ Nexus

THE EGYPTIAN SPHINX HOLDS MANY SECRETS

IT IS MORE ANCIENT THAN OUR EARLIEST SUMERIC CIVILISATION

..So who was it made by?

Our human origin is still clouded.

Our existence appears quite suddenly out of the skies

Something to do with apes,

Reptilian brains and some starseed.

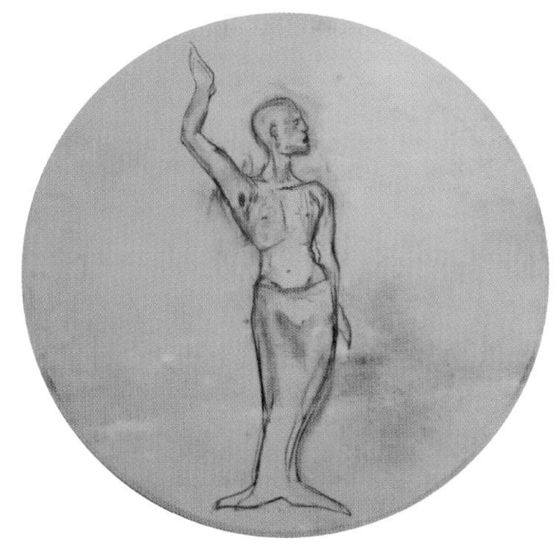

Enlil, Just Left the Waters
Oils on Goatskin Drum ø 50 cm

Cerebellum
Oils on Goatskin Drum ø 75 cm

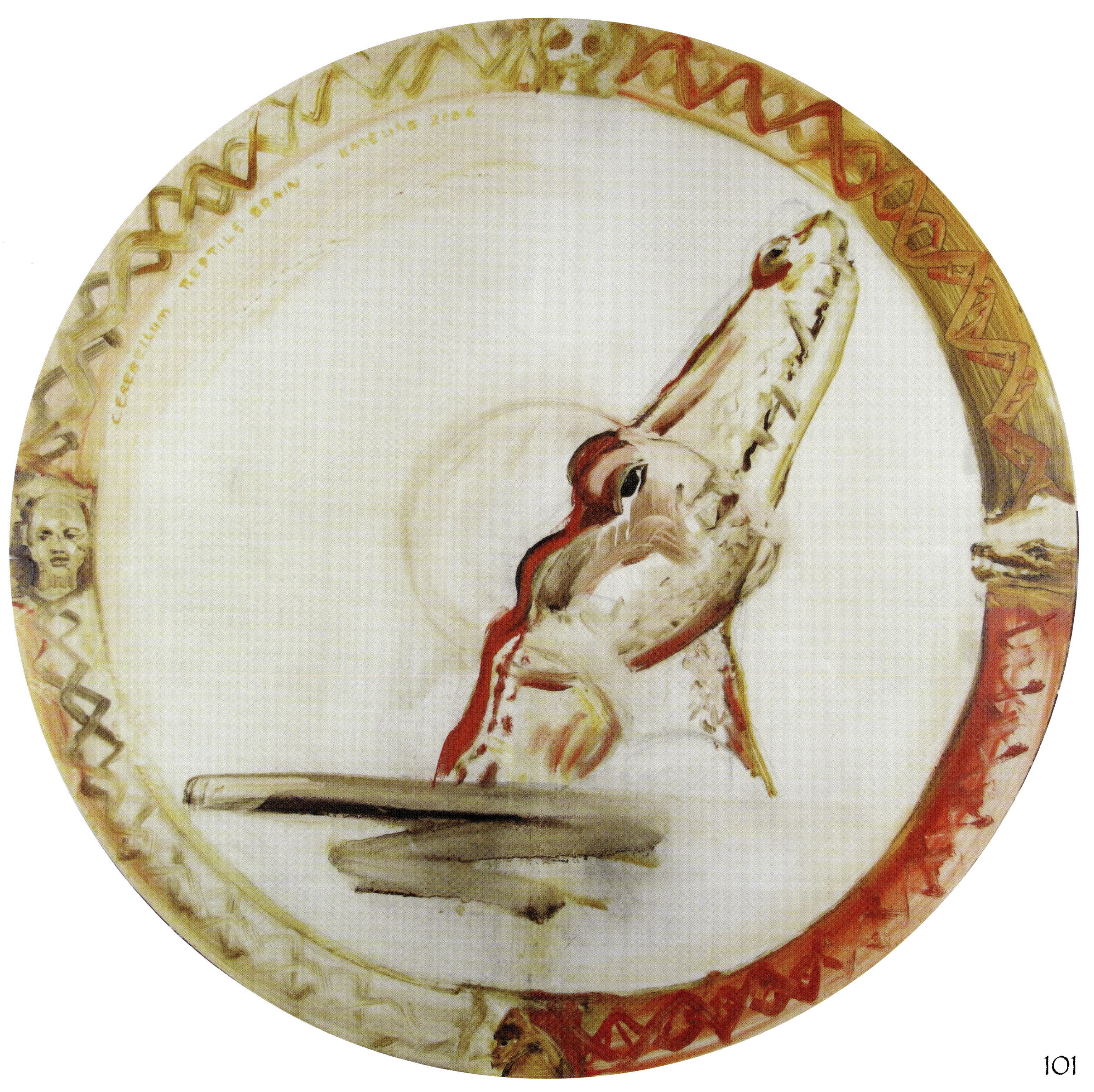
CEREBELLUM REPTILE BRAIN - KARELIAS 2006

≐ Nexus

THE HIGH CIVILISATION OF ATLANTIS WAS DESTROYED

BY THE LAST POLESHIFT (A 90° ROTATION?)

Gaia fell down to a lower dimension

All religions make record of this disaster

The Ark of Noah sails on top of the Flood

Momentarily we move towards the Galactic Centre.

So the upcoming shift will lift our consciousness..

The High Civilization of Atlantis

Oils on Goatskin Drum ø 75 cm

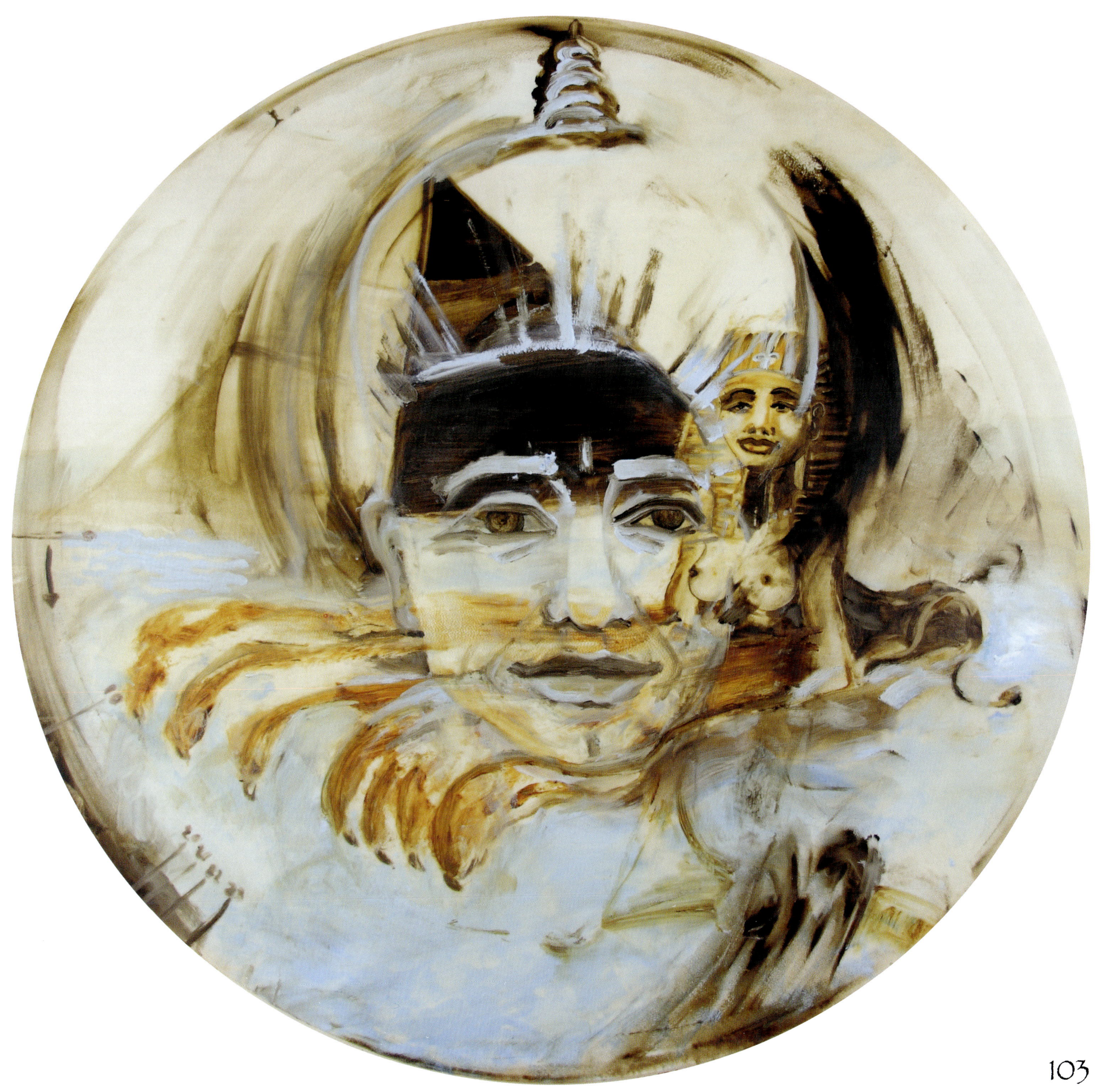

≐ Nexus

Adam & Eve leaving the garden of E.Den, Seth & Osiris in Egypt,

Seven-Macaw in Popul Vuh, Ask and Embla among Norses

All describe how Humanity lost the third eye, which led to a polarised reality.

Buddha came to teach the enlighted state on Earth.

Jesus came to awaken Christ-consciousness within ourselves.

Quetzalcoatl came to navigate through duality.

THE KEY IS WHITHIN, THE HARDEST PLACE TO LOOK

Myths Reveal the History of Conciousness
Oils on Goatskin Drum ø 75 cm

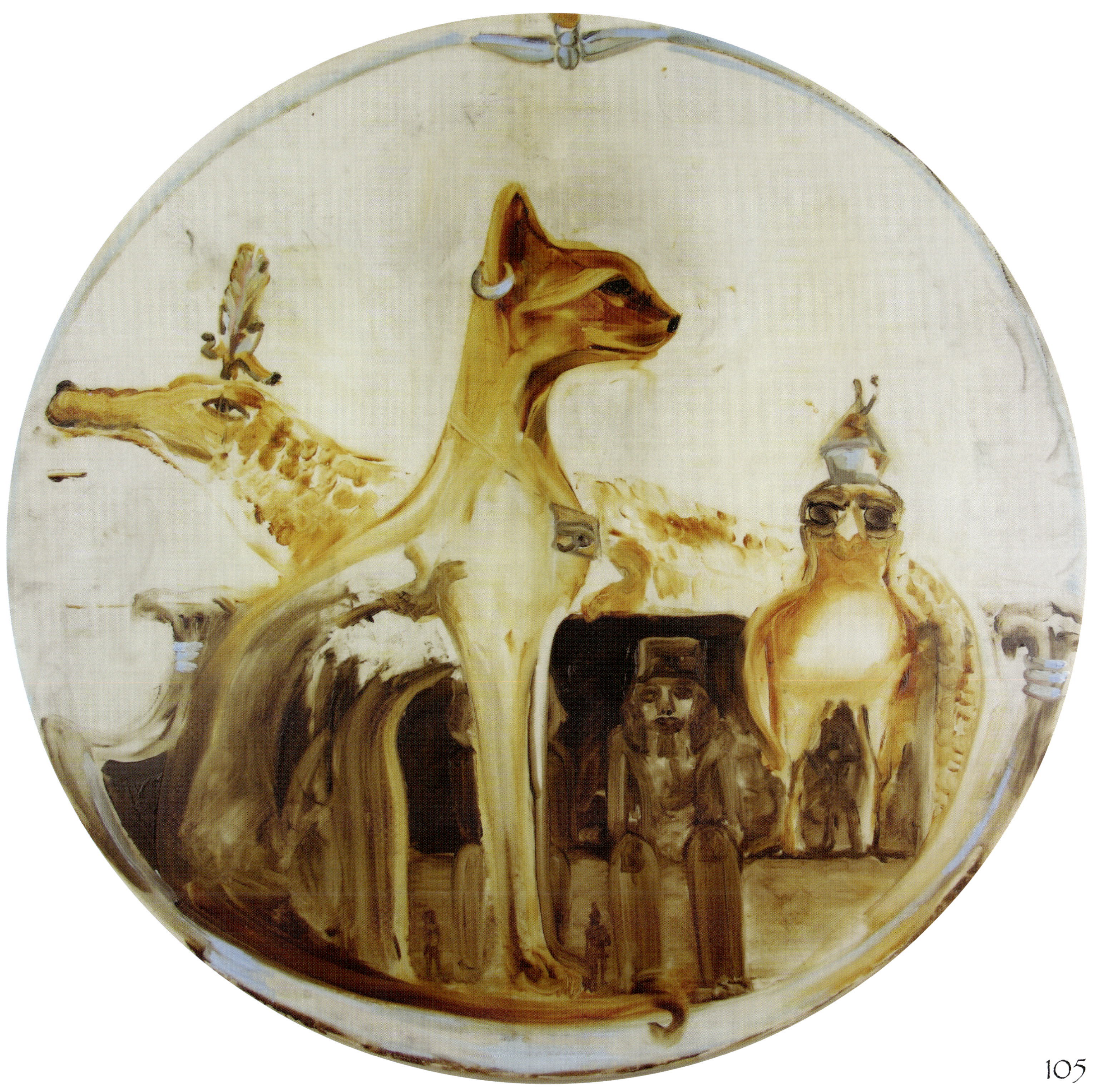

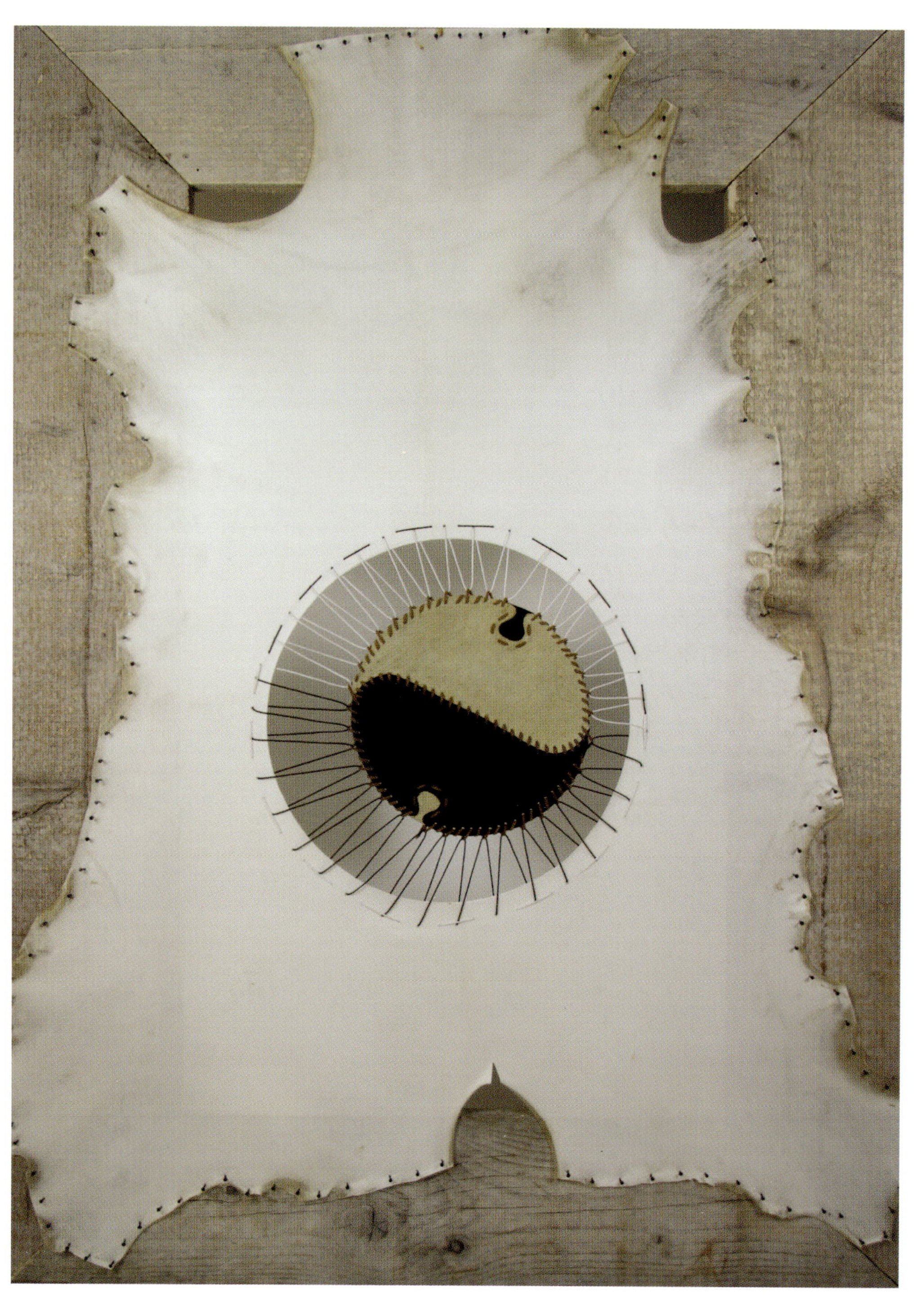

Gaia Envolving into next Dimension
Leather and Wood with Goatskin 135 cm by 95 cm

Gaia

Robbed, misused, even our negative thoughts wear Gaia down,
shakes, quakes are natural responses of a wounded body.
Precious Gaia is meant for a higher purpose
Cows, Pigs and Chickens have decided 'en masse'
rather to die of a nasty disease than live a lousy life,
with only purpose to be the meat on our plates.
What's a life without any sunshine?

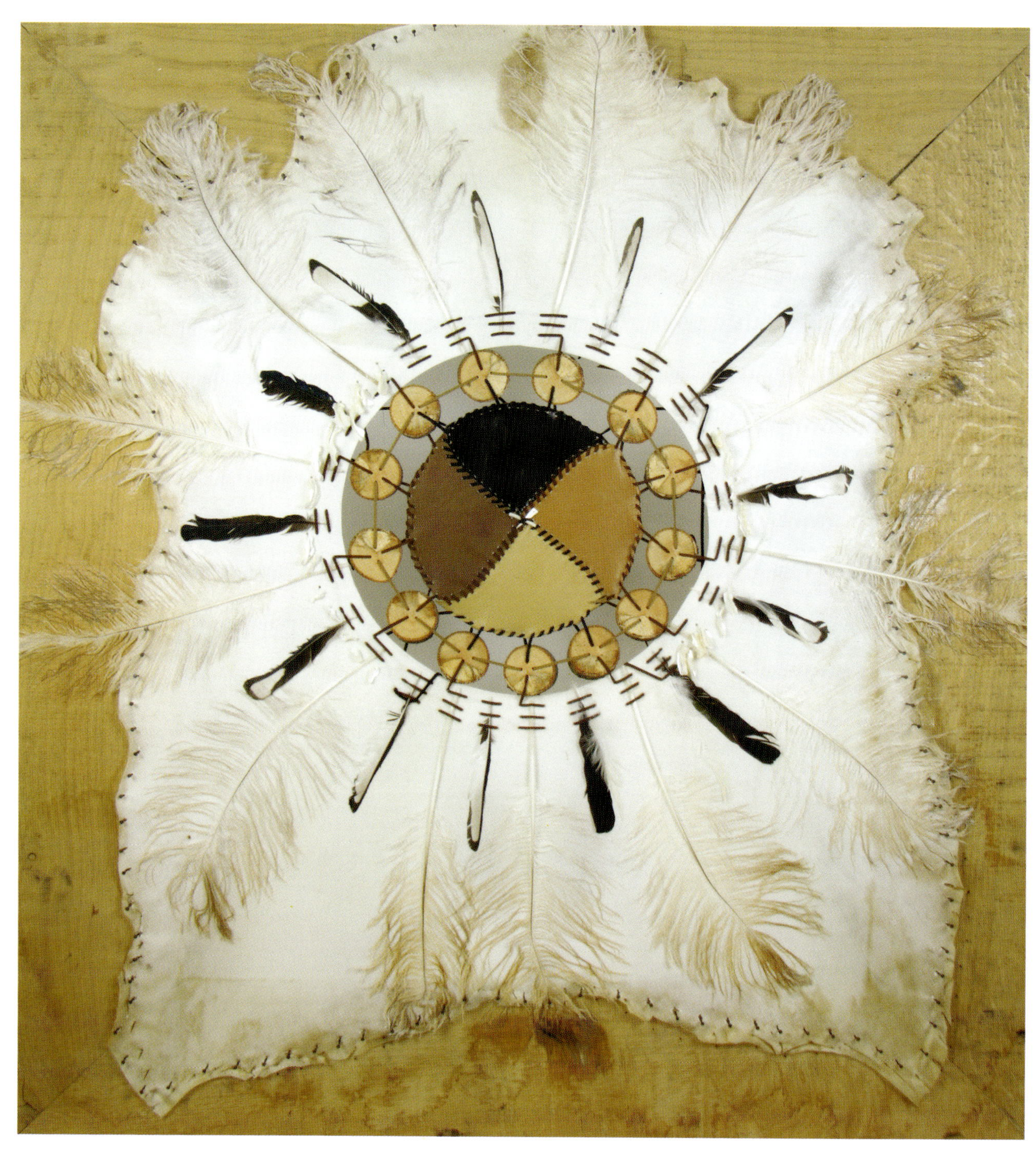

The Axes: Equinox & Solstitia

Feathers, Leather and Wood with Goatskin 120 cm by 110 cm

Gaia

Celtic Cross, Medicine Wheel, World Tree, Tree of Life,

all describe the energies axes.

The aboriginals are one of few,

still in touch with Gaia's energies.

To strengthen themselves, they walk along the telluric lines.

Churches were built on the Earth's strongest power places.

We used to build our homes in line with them.

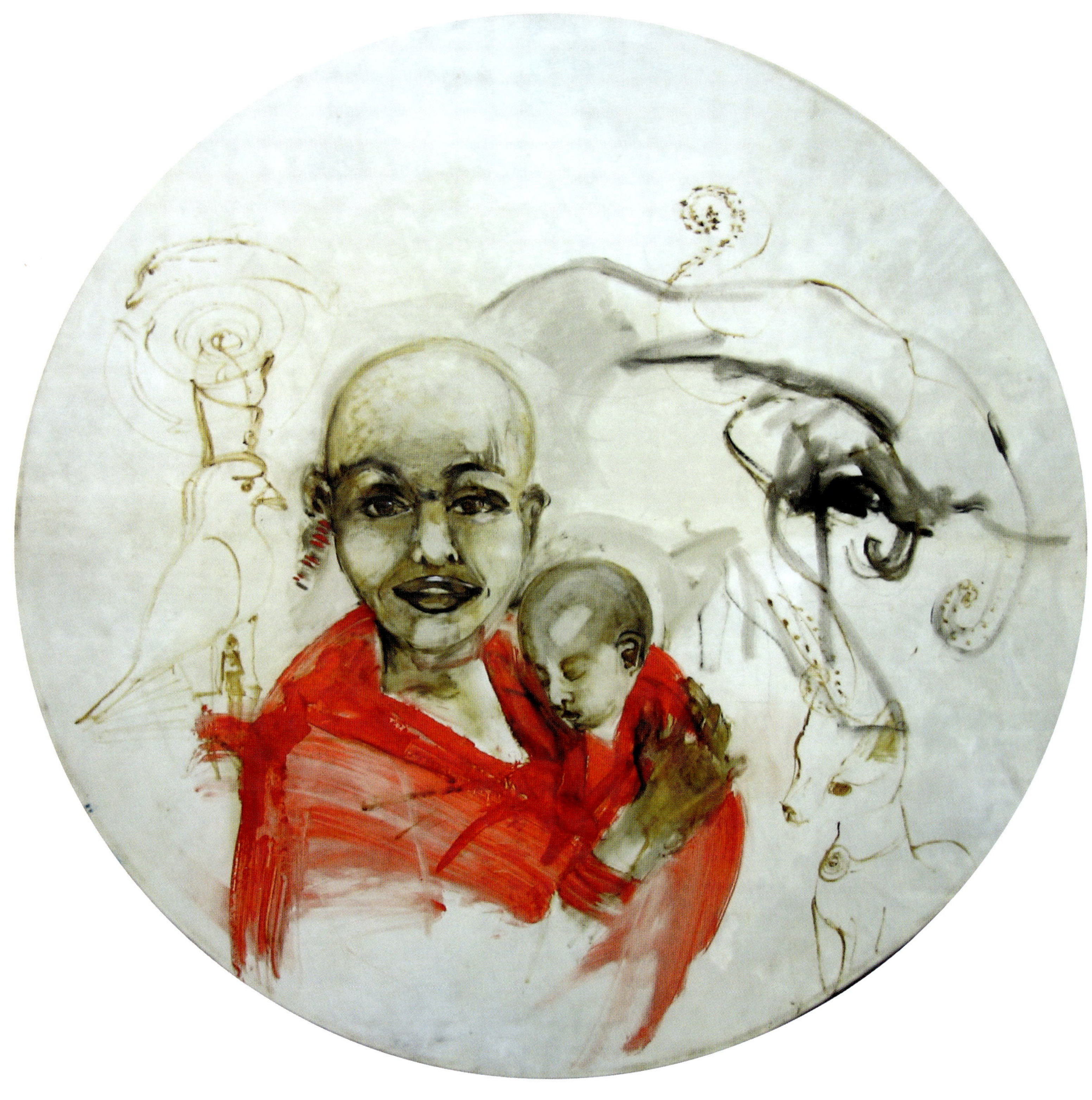

Gaia
Oils on Goatskin Drum ø 95 cm

Gaia

Mother Gaia wants us to listen carefully and remember:

Her heartbeat, that Gaia's inner core corresponds to the stars

Like her children: *Her Cats hiss to the dog.*

Her Dolphins dance with a dwarf.

All Birds flap about the sisters' heads.

The Bear loves milk from far and away.

So that's why Sauriërs look like dragons.

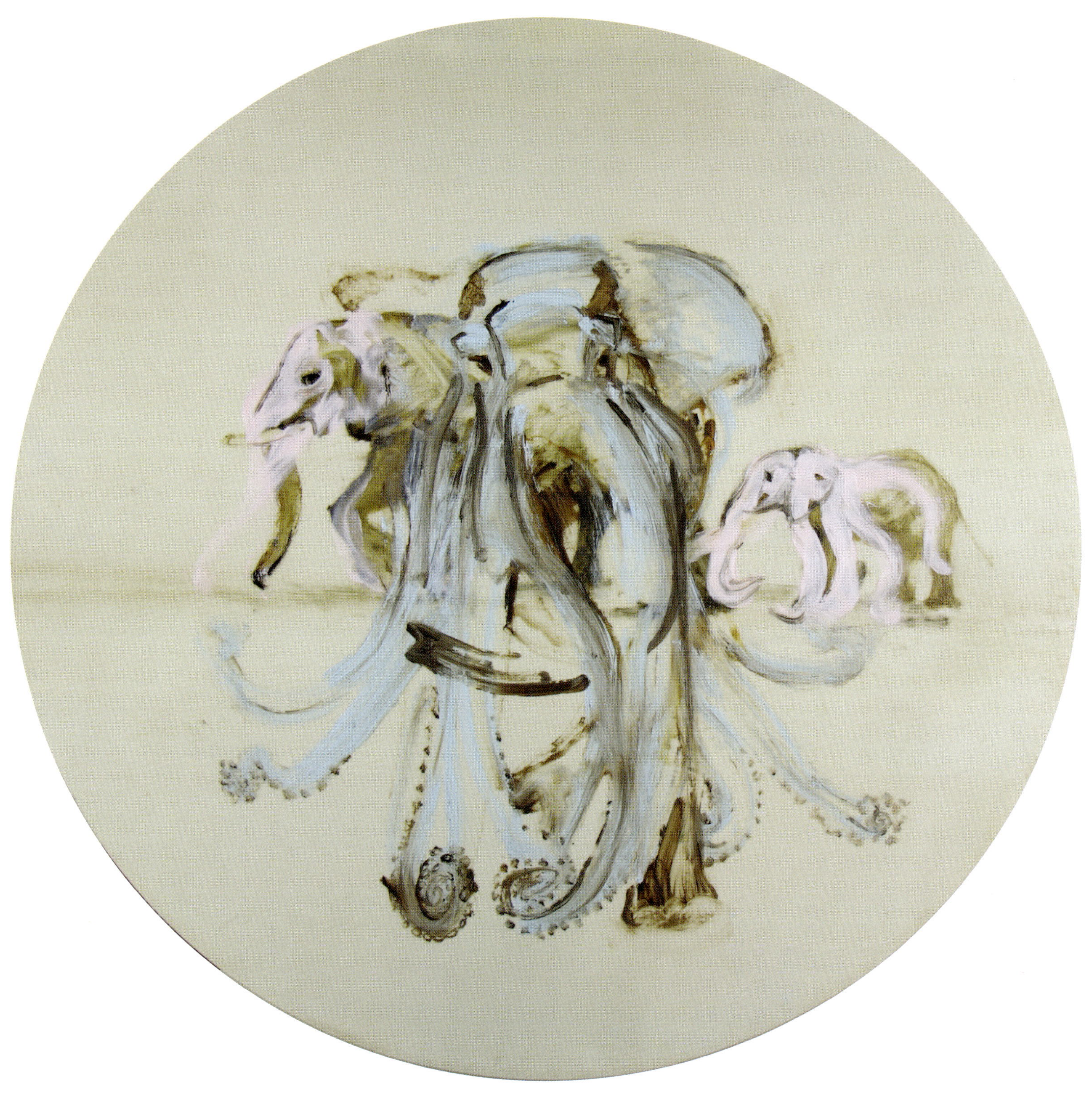

Animal Energies

Oils on Goatskin Drum ø 75 cm

Lunar Dragon, bronze

Gaia

All Animals, Plants and Stones are beholders of specific energies,

all a piece of the puzzle to remind us what humanness means.

The joys and responsibilities as keepers of planet Earth.

We must nurture the planet, thank her, help her to feel her best.

So she can mediate, via the Sun, other cosmic energies back to us.

Because all life on Earth represents Star consciousness,

Everything is connected therefore:

All There Is Matters Equally.

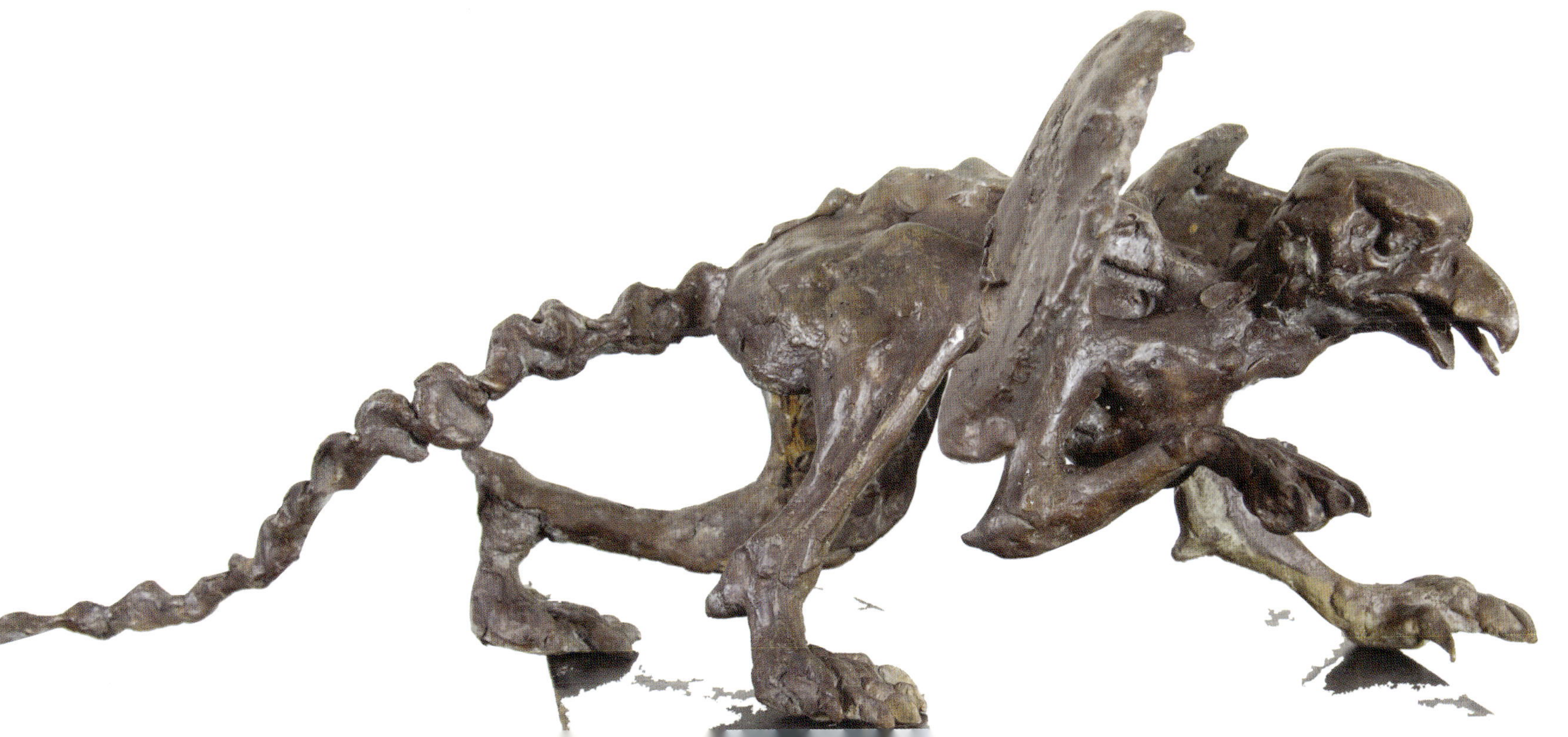

Cosmos

The Tree of Life is rooted in the Underworld,

with branches holding up the Heavens in the sky.

An invisible web of energy lines

which exists at different dimensions in the universe.

Changes in energies take place simultaneously

at turning point of cycles in the cosmic plan,

Like the health or sickness in your own body.

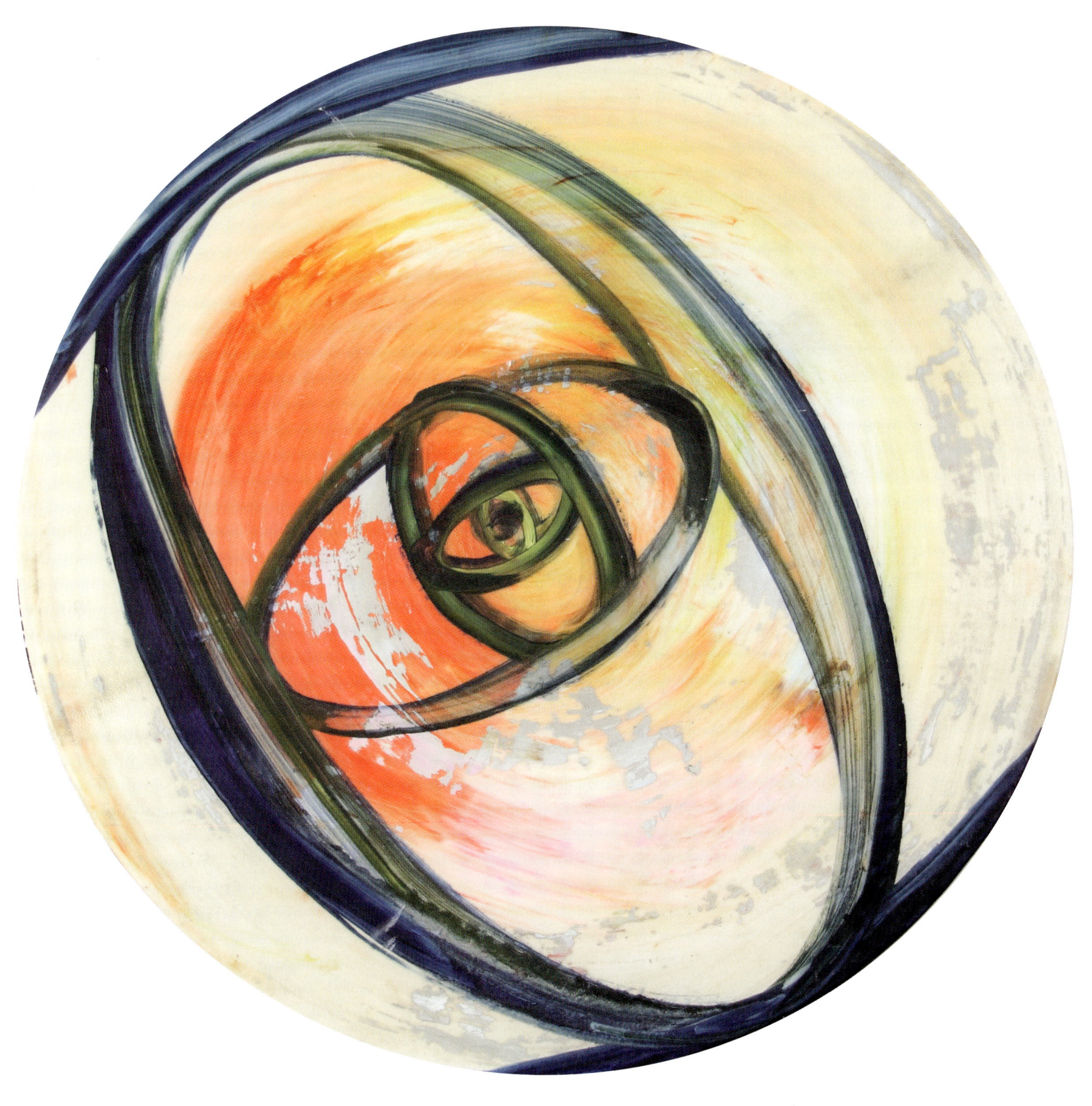

Cosmic Shifting
Oils on Goatskin Drum ø 50 cm

Cosmos

EVERYTHING IS RELATED.

THEREFORE SPIRIT AND MATTER

CAN NO LONGER BE SEPARATED.

Our cosmos is alive and pulsating.

Unified through the energies

of divine creation

described by the Maya Calendar.

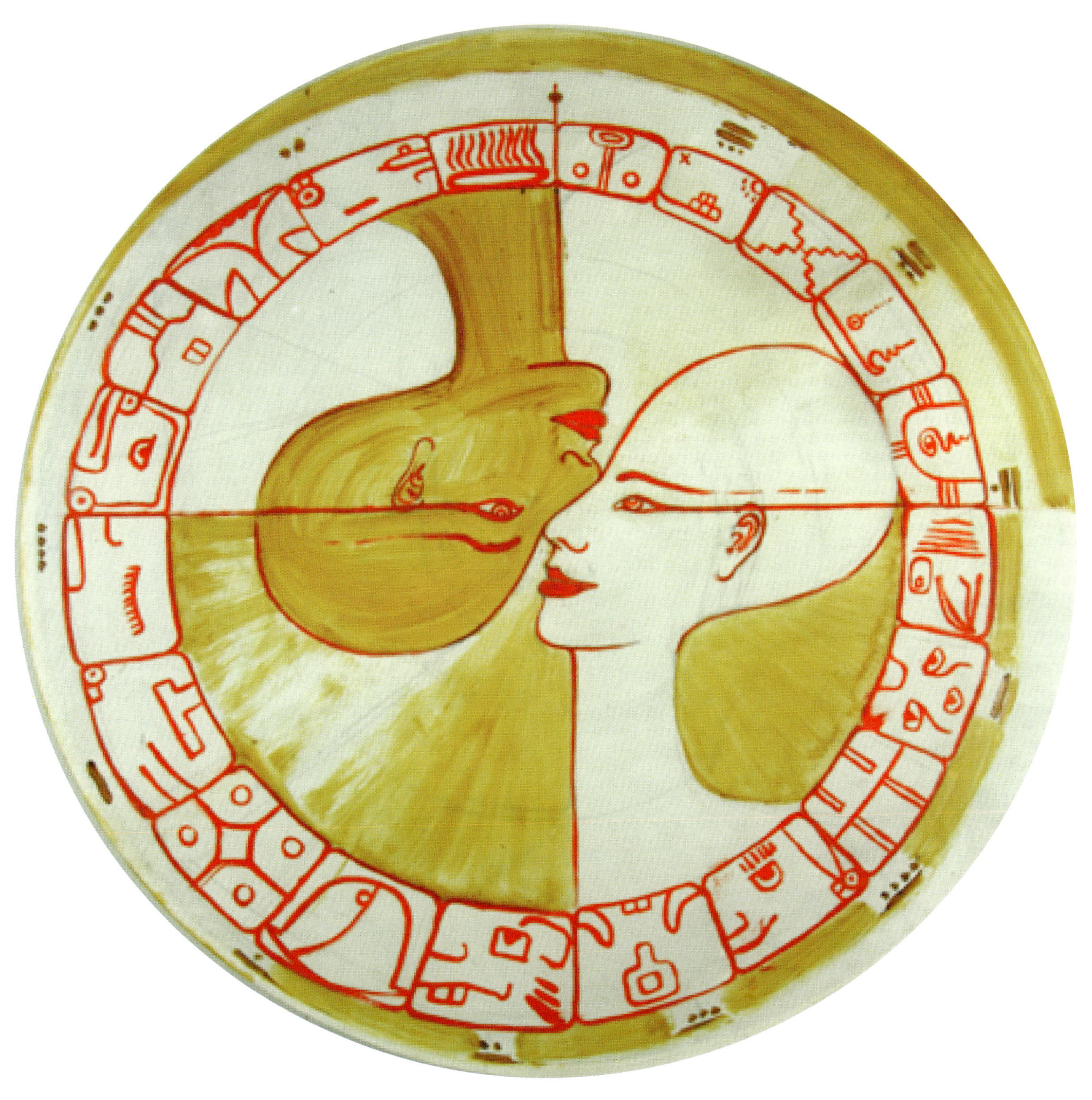

Tzolkin Round
Oils on Goatskin Drum ø 70 cm

Cosmos

So all your actions matter.

When you move with love, humor or compassion

you harmonise with light, lucky you.

Otherwise, life backlashes at you.

It is the power of the shadow.

Light and darkness represent the Code of the Cosmos

Time to stop the unawareness and act with harmony in mind

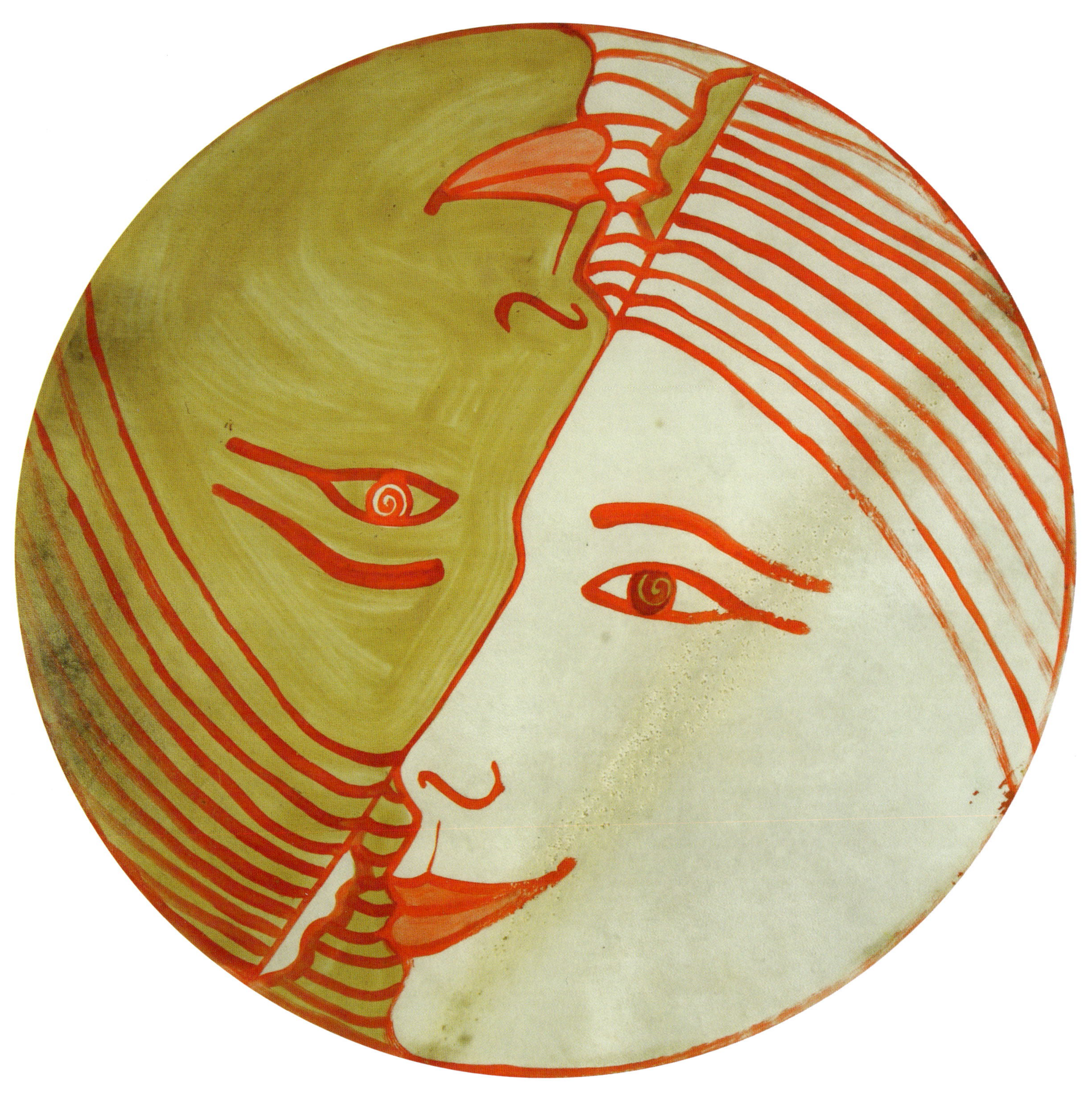

Action & Reaction
Oils on Goatskin Drum ø 50 cm

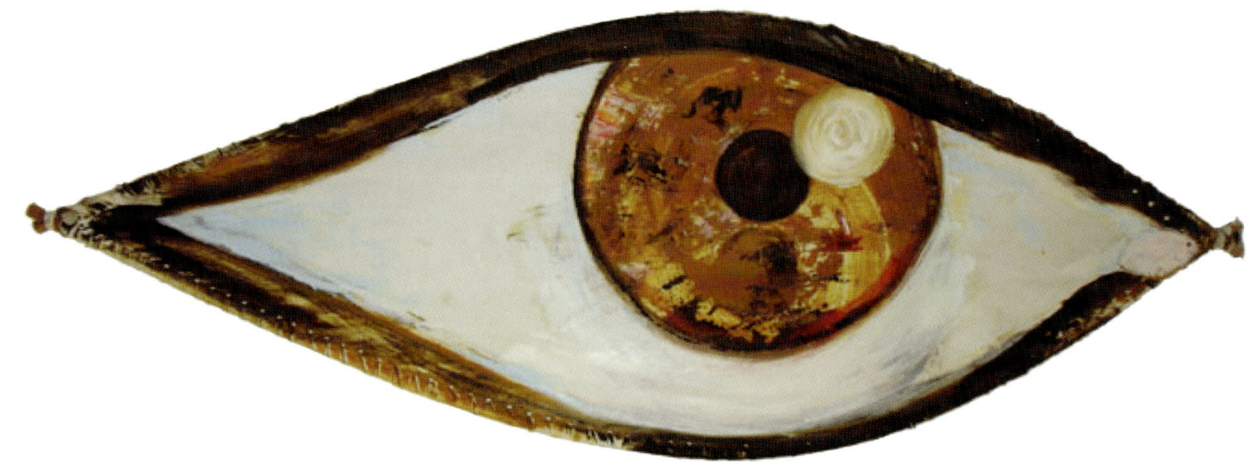

⩸ Cosmos

All souls have gathered on Earth to be part of the SHIFT,

a new light is shining,

a new sound is vibrating,

a new time is approaching,

a new humanity is evolving,

and a new planet is appearing.

WATCH THE NEW CHILDREN LEAD THE WAY.

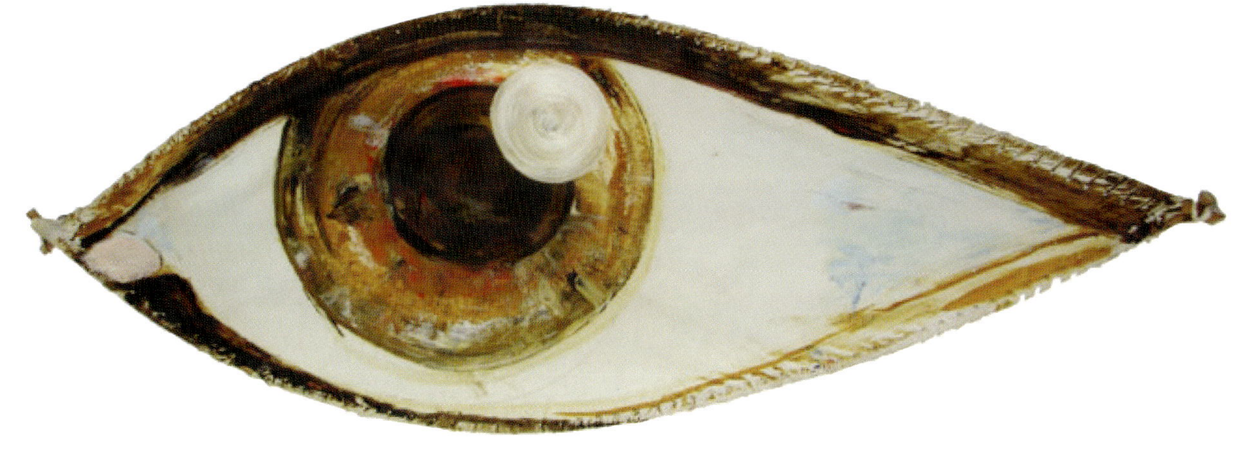

Watch & See

Oils on Goatskin Shields 150 cm by 55 cm

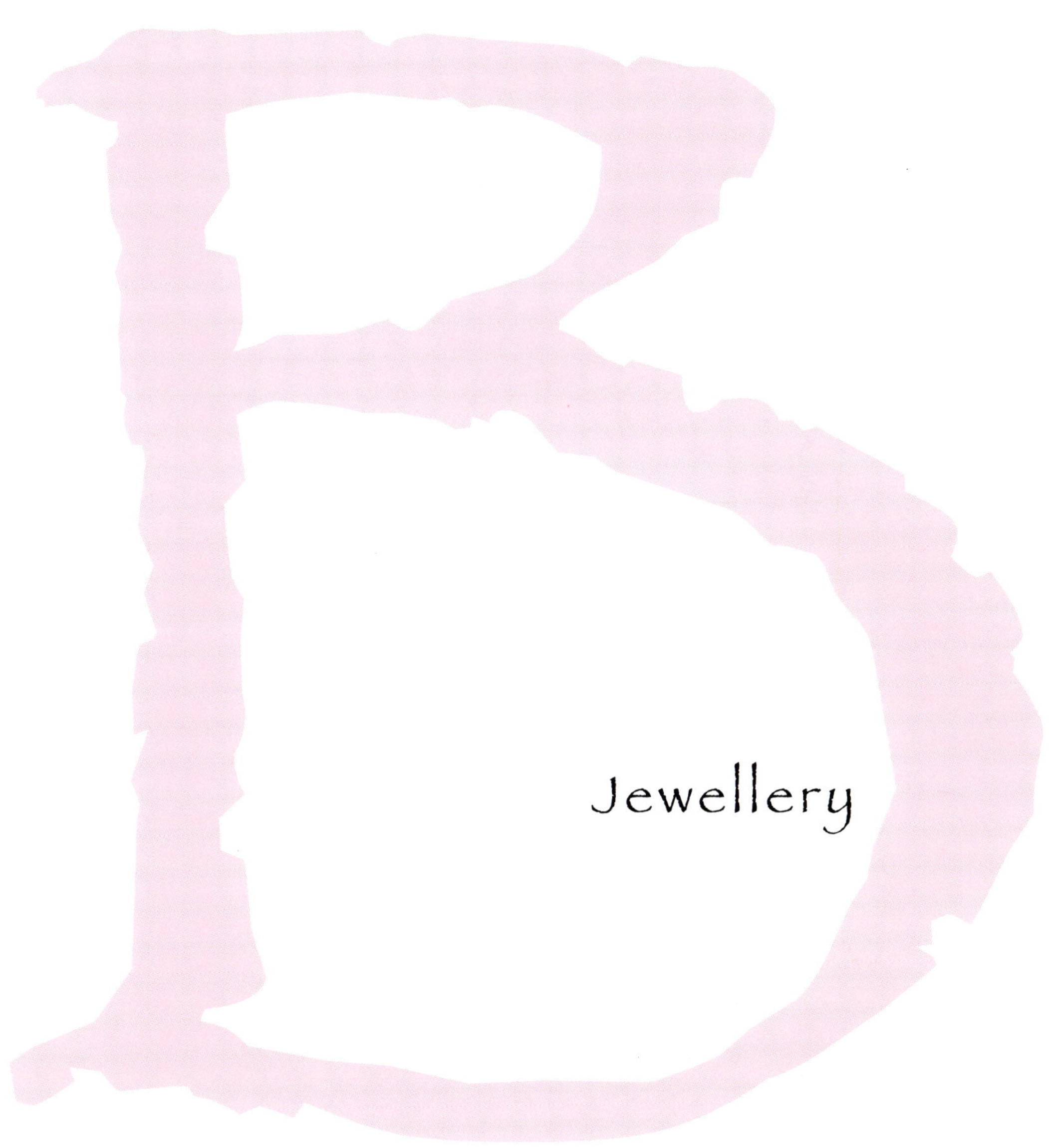

Jewellery

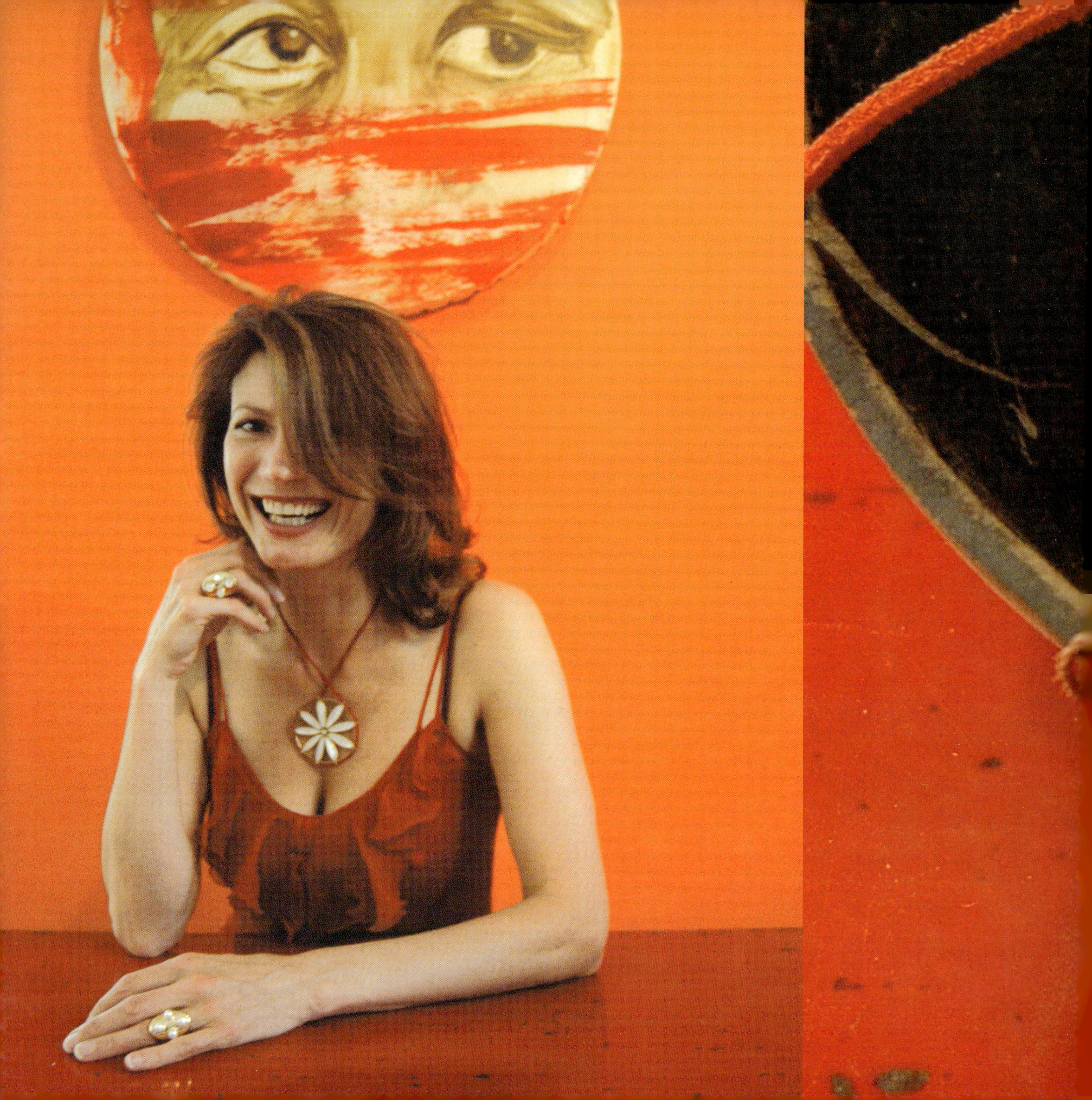

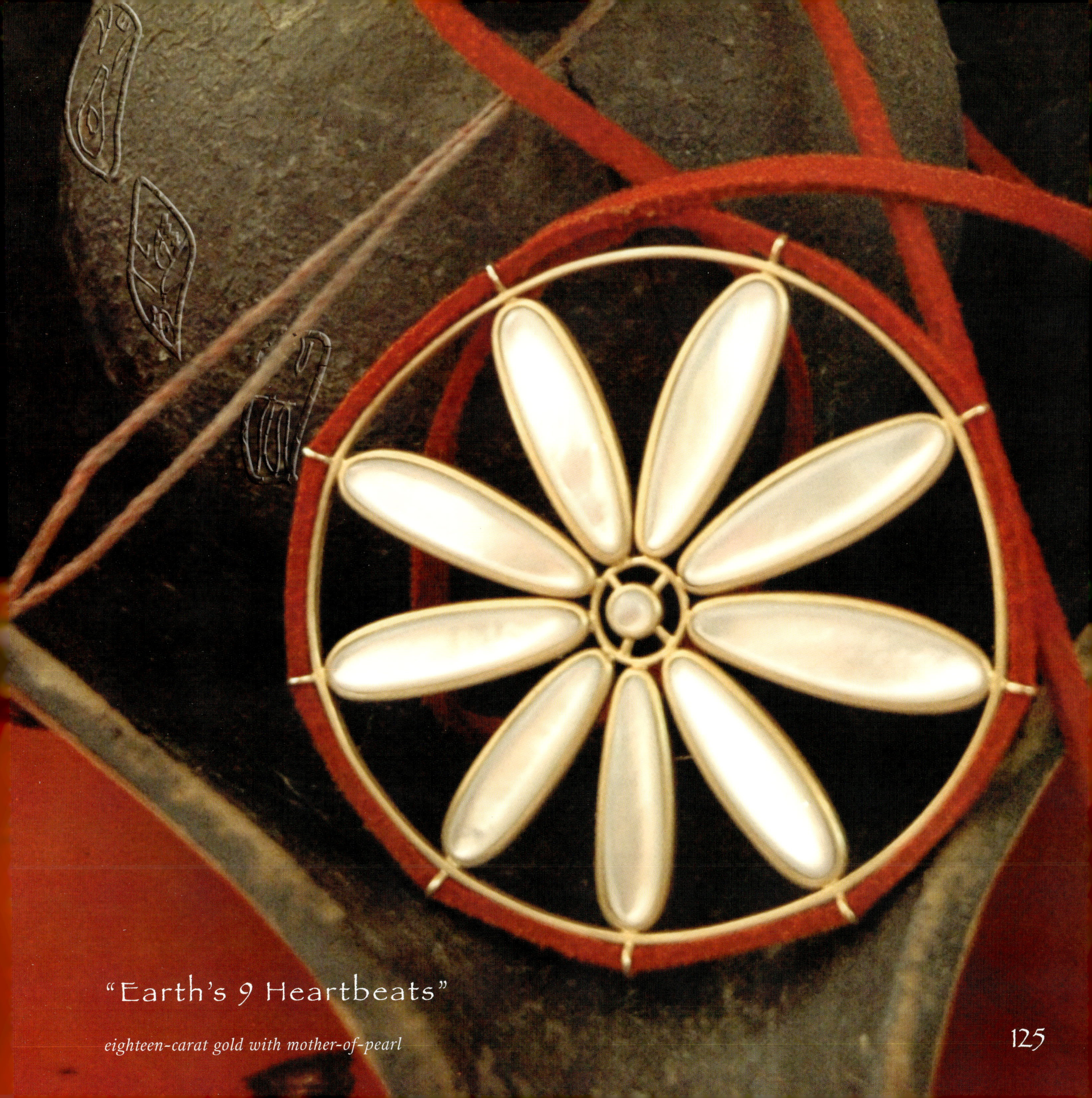

"Earth's 9 Heartbeats"

eighteen-carat gold with mother-of-pearl

"13 Faces of Creation" and "Flower of Life"

white gold with black and white pearls

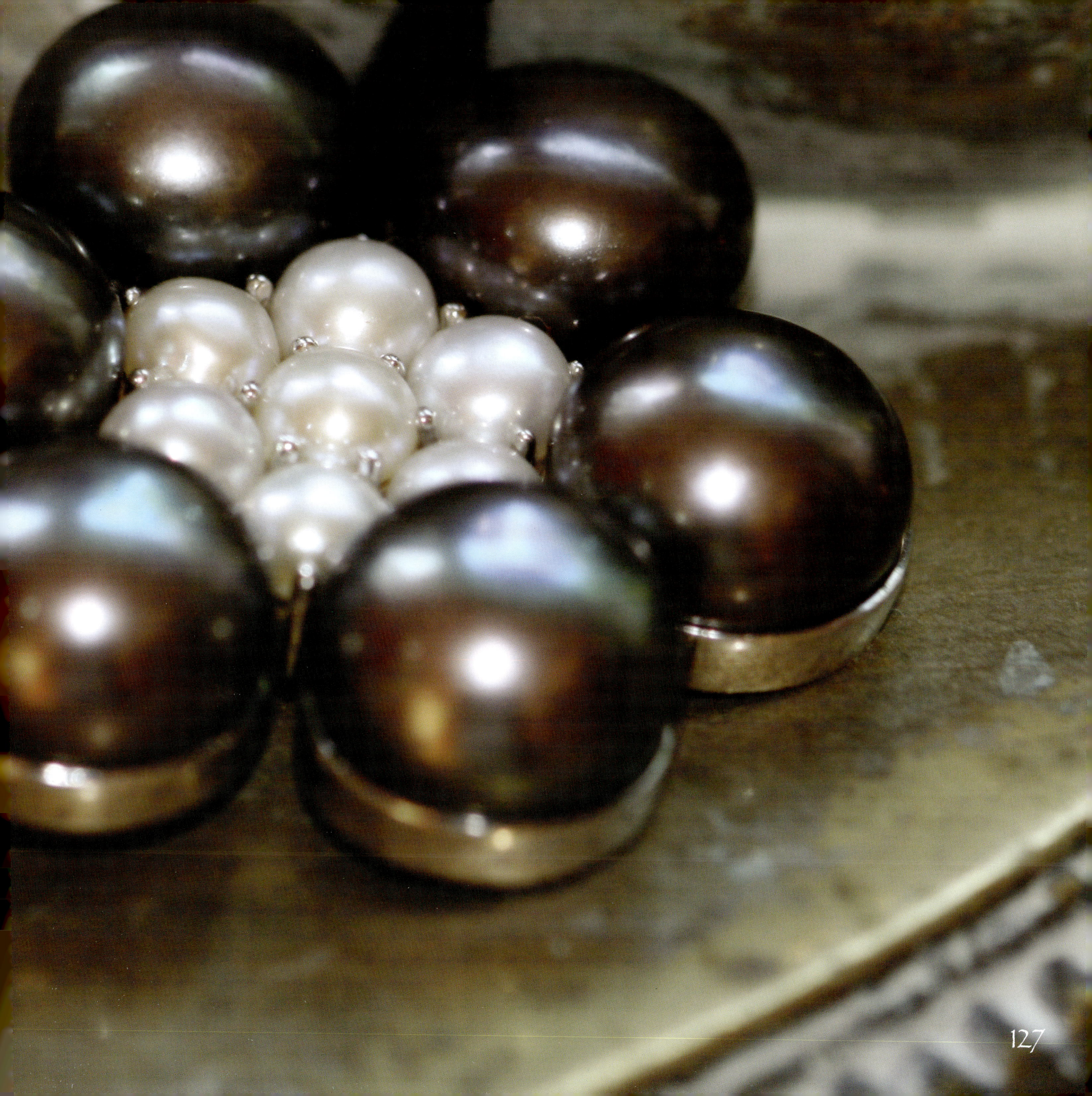

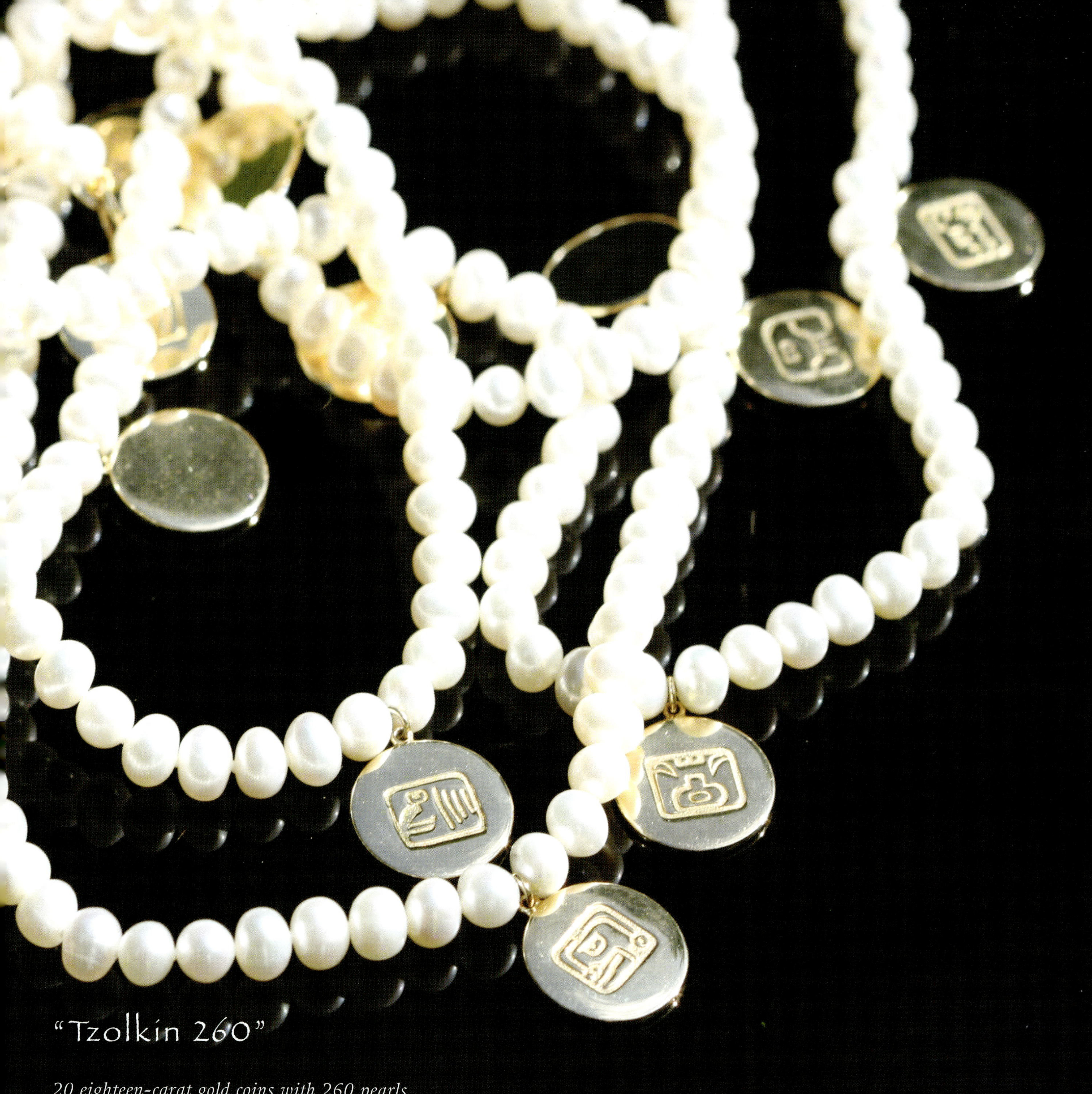

"Tzolkin 260"

20 eighteen-carat gold coins with 260 pearls

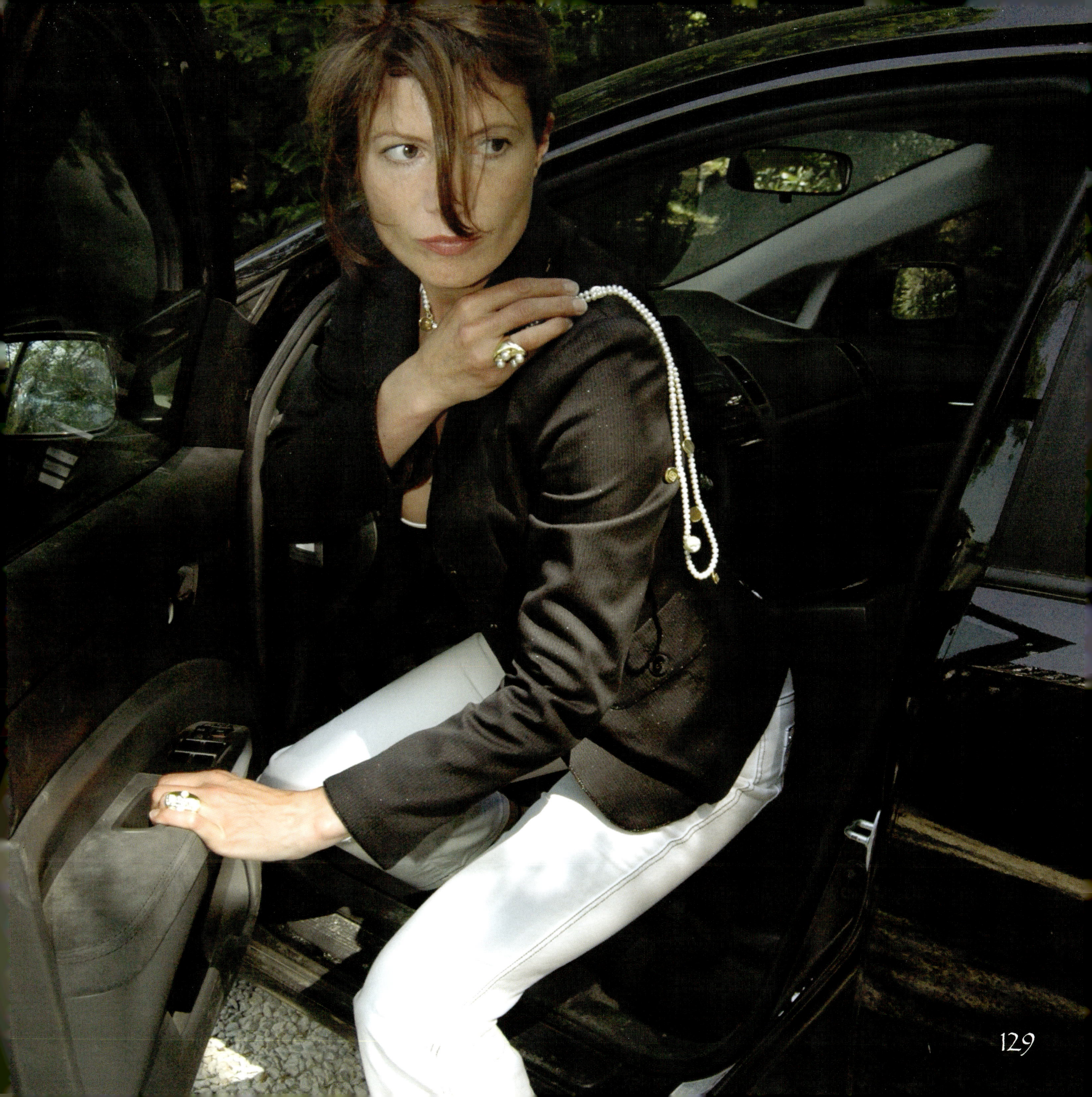

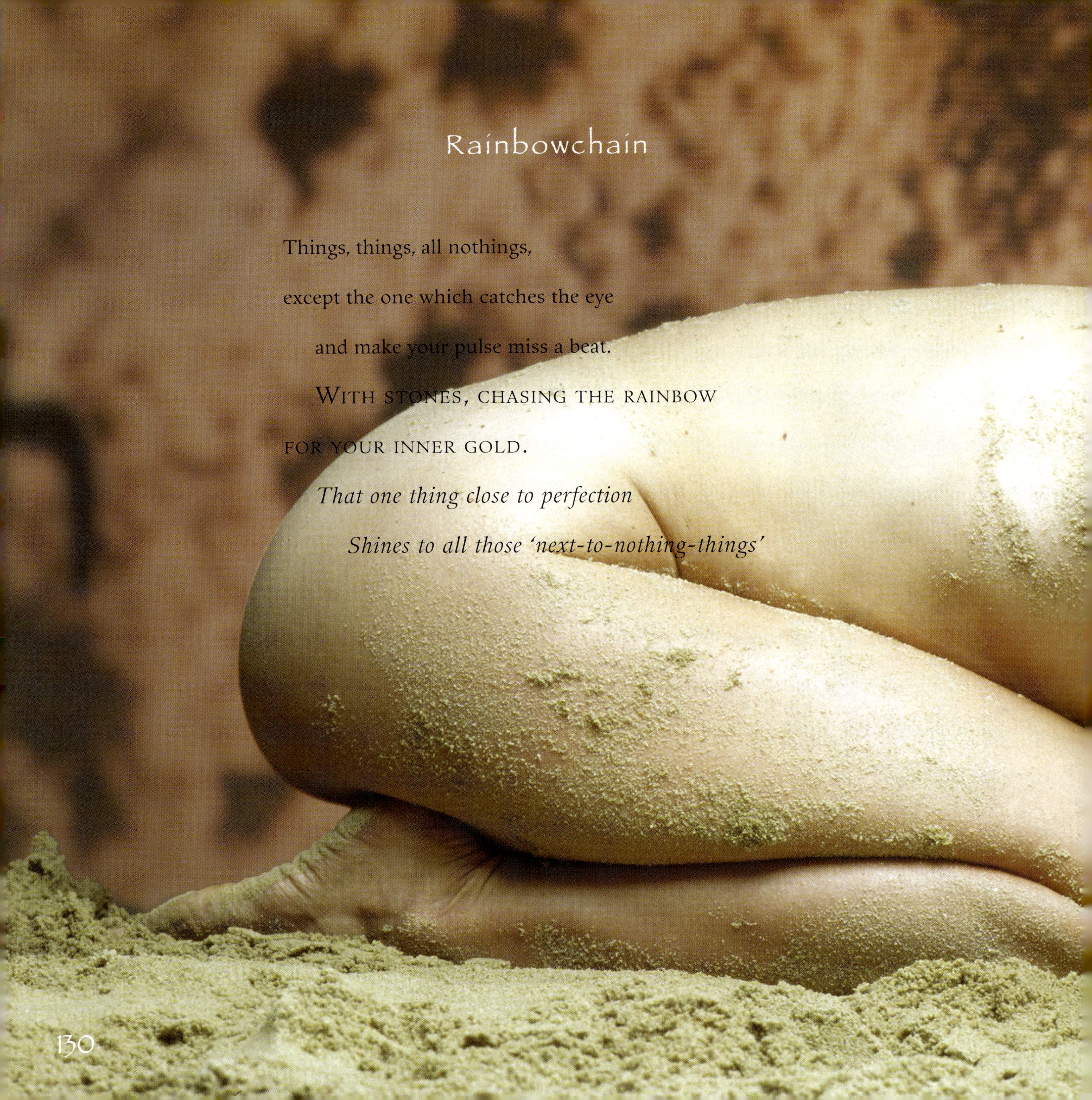

Rainbowchain

Things, things, all nothings,
except the one which catches the eye
and make your pulse miss a beat.
WITH STONES, CHASING THE RAINBOW
FOR YOUR INNER GOLD.
That one thing close to perfection
Shines to all those 'next-to-nothing-things'

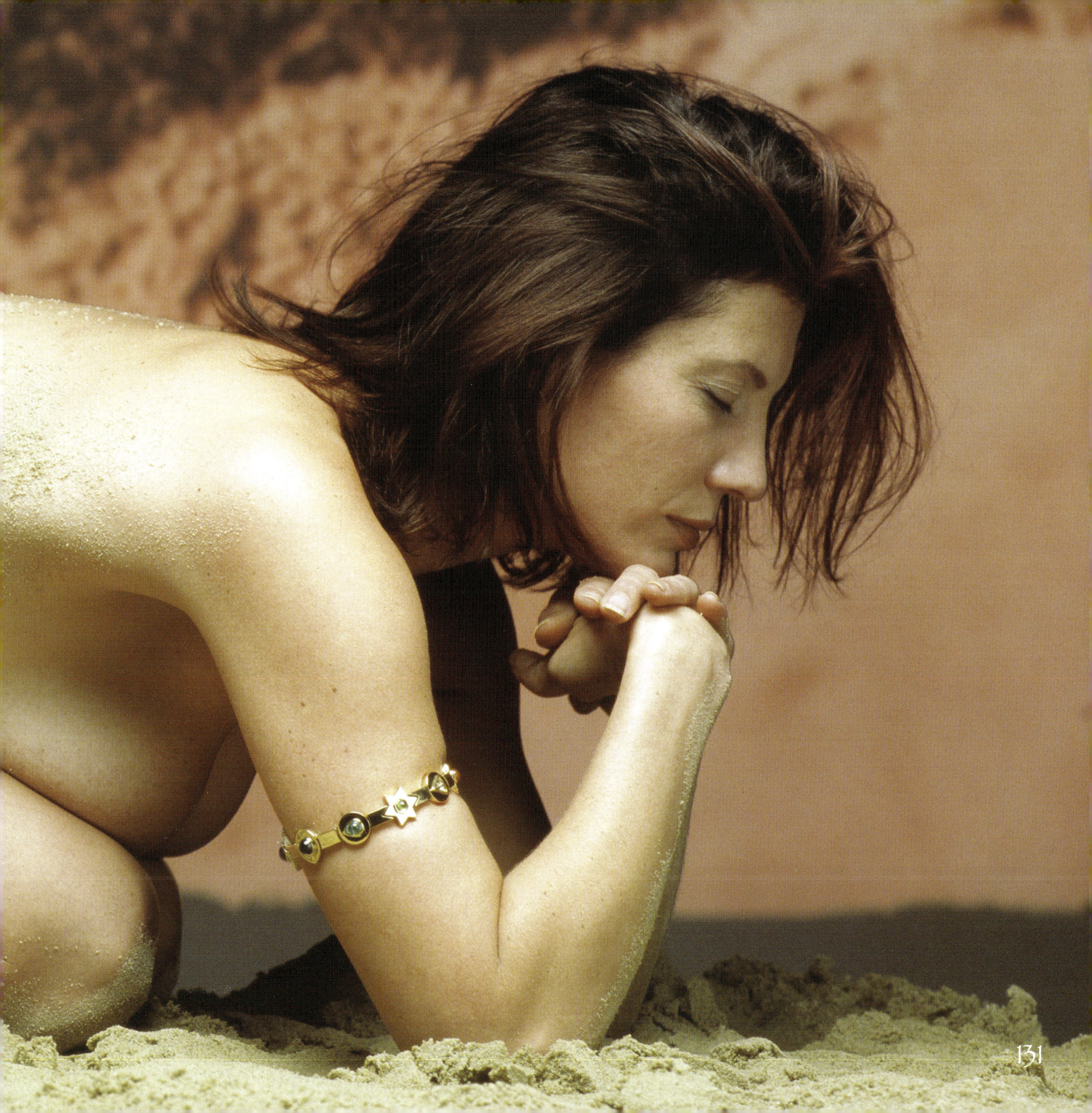

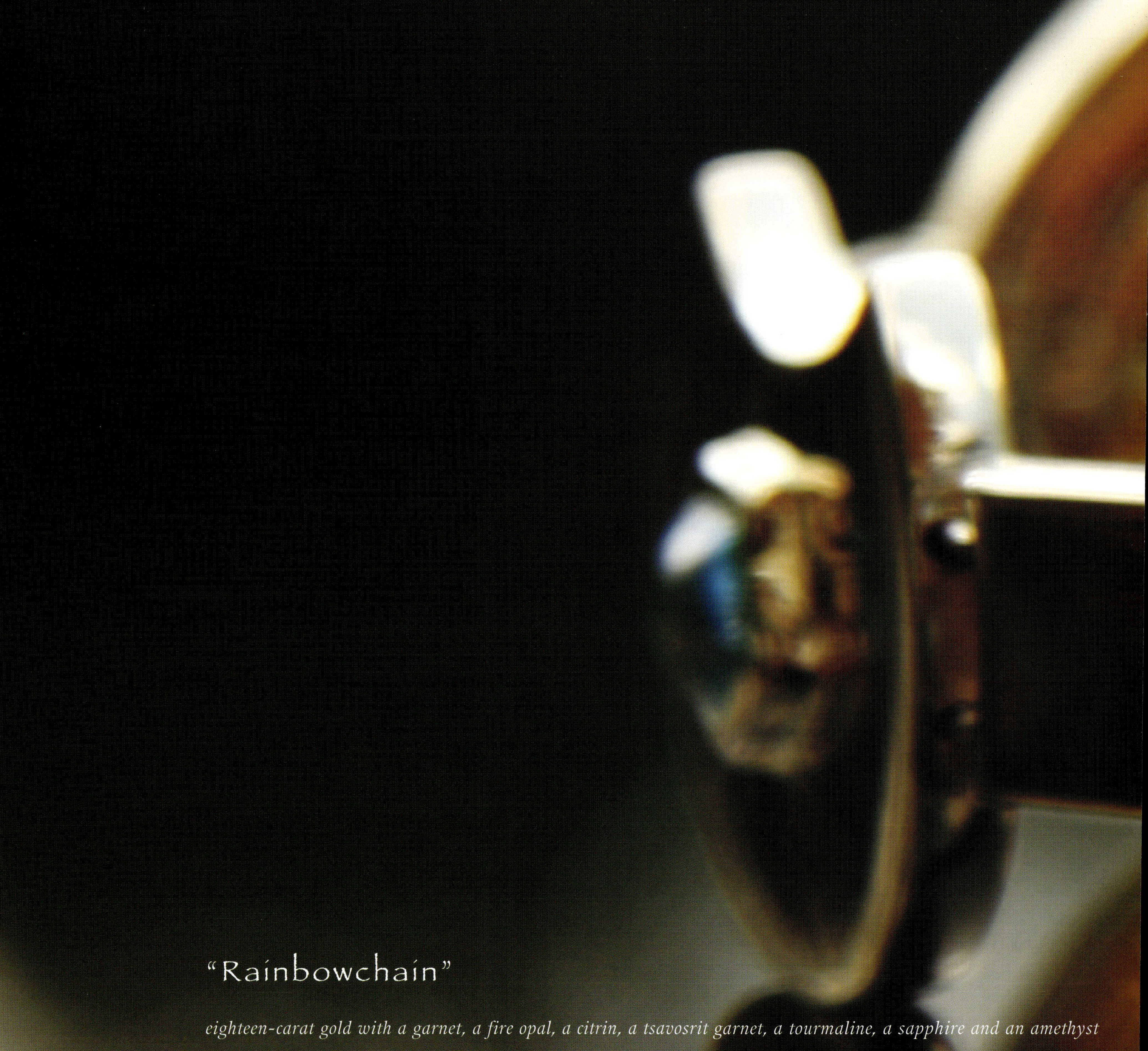

“Rainbowchain”

eighteen-carat gold with a garnet, a fire opal, a citrin, a tsavosrit garnet, a tourmaline, a sapphire and an amethyst

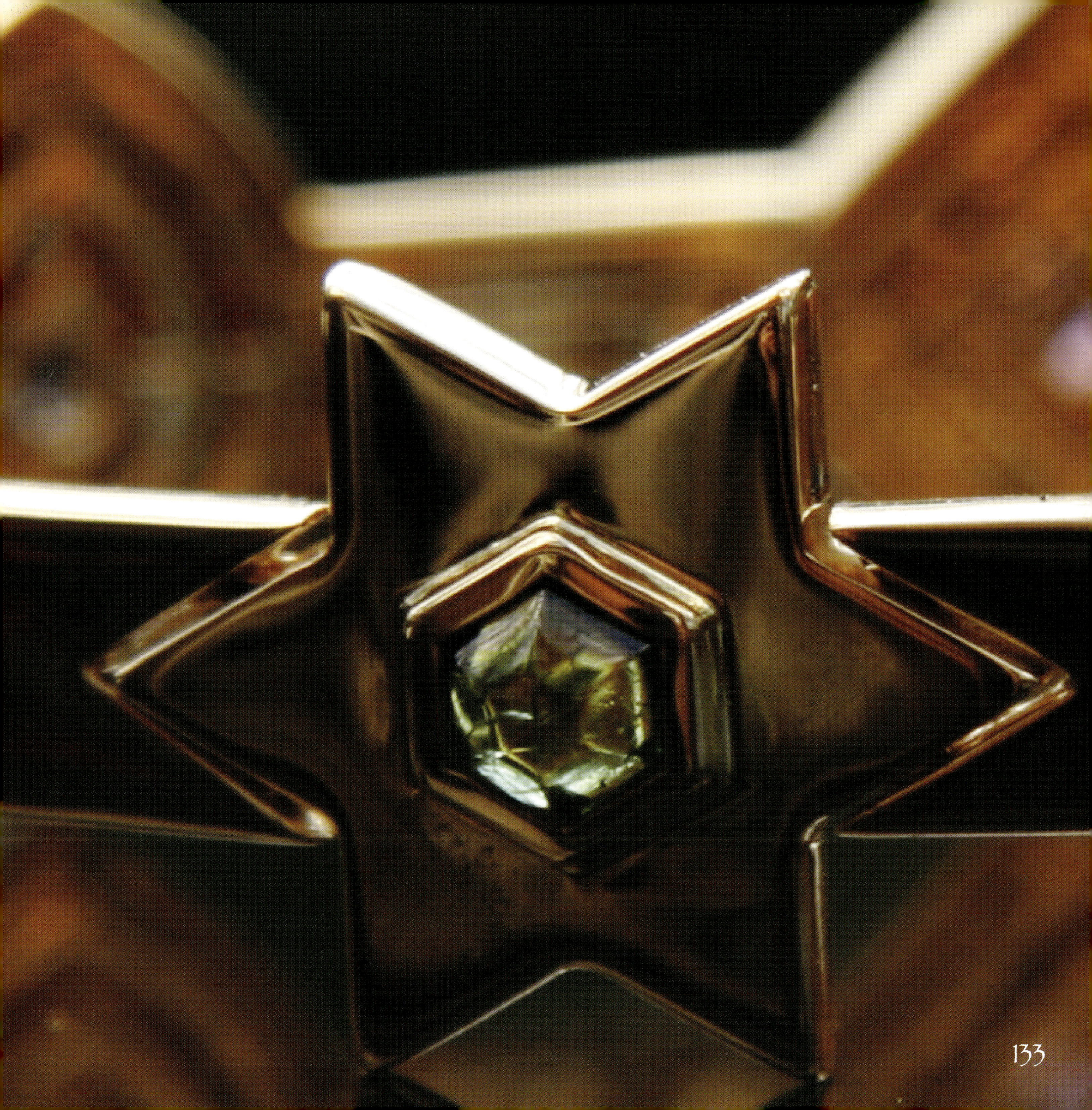

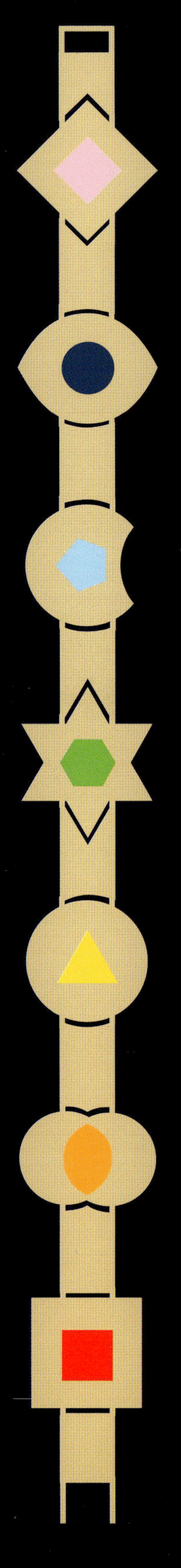

Red is about ONESELF
red is colour number one
red is the beginning
the foundation of things

first, you crawl on all fours
until the day
you put your legs under your body
lift your hands of the ground
and rise yourself up into the air
and there you are:
standing on your own feet
by yourself
in the red soil
alone
good for you, toddler!

now, years later
it is still a big thing
sometimes we need to rest
on someone else for a while
until we strengthen
and can stand by ourselves again
another may need others
relying heavily on him
to get the energy
to stand straight
...whatever...

confidence in
yourself alone
puts you on
your own feet
and when you feel self-assured
and at ease with life,
your feet will make
a perfect square by itself:
SQUARE ONE, that's a big step!

a golden square is the ground shape
the red stone, a garnet
represents the earth
a big thing, the earth,
to give you confidence
it is always there for you,
just under your feet

the gold is heart shaped
for Love is a matter of the heart
at the centre a perfect round stone
to express the pureness of the
feeling
all round things like movement
so stir it up now and then
the stone is a fire opal
which is known for its ability
to awaken deep feelings

Orange is about LOVE
orange is the second colour of
the rainbow

it is about 2
it is about you and your Love
it is the Love you'll give
a lifetime long
that is like........ forever!
that is real Love

Yellow is about
LAUGHING
yellow is third in line
it is about sunshine

yellow is you laughing out loud
the sound of sheer happiness
makes troubles disappear

to hear your roaring laughter
is what life loves best
that is why it favors you

yellow is also the smile of
self mockery
which makes a disaster
so much more bearable

yellow is the sun who makes
a shimmer shine
but the sparkle of life is only seen
by the twinkling eye

the citrin stone is triangular
which symbolises fire
children full of life
and people who laugh a lot
have that fire thing
going for them
keep it burning
and glow on please

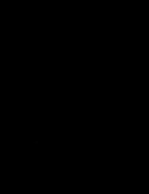

Green is about
HARMONY

green is the colour of a tree
green is growth & peace
active and passiveness
perfectly balanced
all in one thing
a tree

green is about harmony
balancing the upward
and downward forces
what do you take ?
what do you give back ?
what do you lose ?
what do you get in return ?

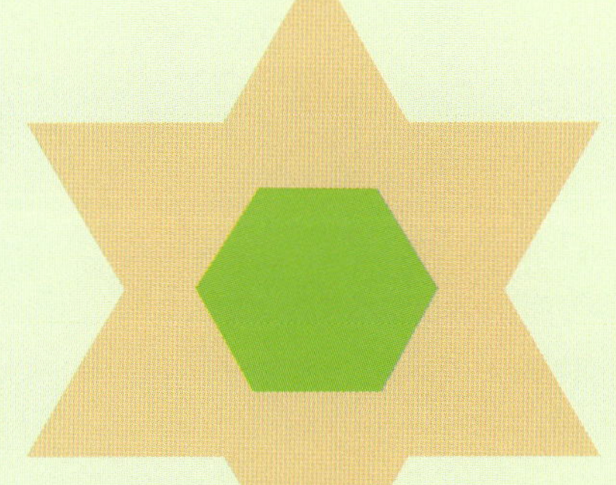

growth & peace
in a tsavosrit garnet

green trees help you to rest
and restore your energy levels
green trees help you grow over it
and get rid of the pain
like falling leaves
just like that, if you let them

and then, like spring
you turn up new
a little different, a little wiser
but you hardly believe your eyes
because the world has changed too
it is a new place to explore
and start all over again
that is weird, but wonderful

Blue is about FREEDOM

blue is the colour
of the clear sky
that faraway colour
which gives you
all the time you need

blue is the colour
of the endless sea
that's full of water
which gives you
all the space you need

water can change form
like it is no~thing
just to help you
to move freely and
to do what you dream of

water runs to you
to lessen your thirst
or when your thoughts
turn into a knot
rain will freshen your brain

and when water feels
you are out of synch
it lingers on
and turns into ice
to cool you down

and when water thinks
you've had enough sunshine
it rises up
and becomes a cloud
to dim the light

water is so adaptive
it interacts with all sorts alike
the cycle of water is caught in
a golden circle with a tourmaline
shaped as a drop of water

Indigo is about INSIGHT

Indigo is the 6th colour
the 6th sense
the observation
only seen by you
with its meaning
only clear to you
Indigo is a bird's eye view
from high up in the air

Indigo is a golden eye
with an iris of sapphire

Indigo is the world
next to reality
it is the place
where we all long to be
filled with dreams come true
new knowledge for blissful ideas
there where you instantly
become your ideal

Indigo is the sky at night
with a million stars to look at
talk to and listen to.... what's up
it is also the deep dark sea
where you have to overcome
your fears to get ahead
but is there greater satisfaction
than the power to overcome
oneself?

The last colour violet
is an amethyst check
to lighten

the
Godly
things

like the
coming & going
of us,
human beings

Violet is about FAITH

without faith,
nothing has ever been done

without faith,
nothing really makes sense
(if you really think about it)

without faith,
life fades away

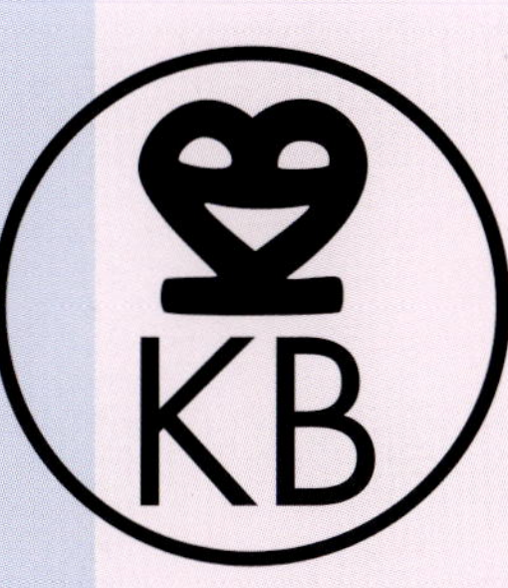

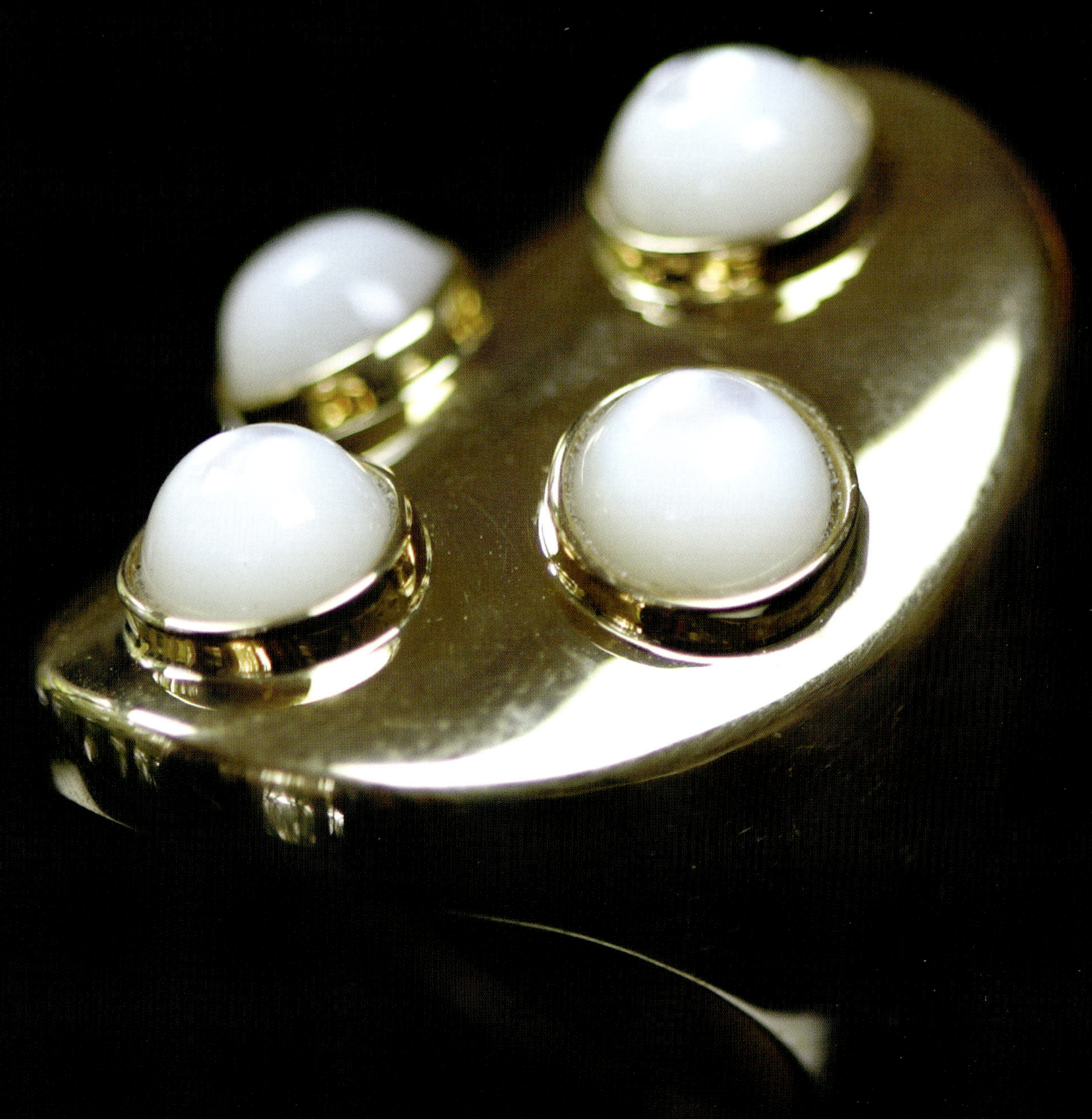

"4 Winds"

eighteen-carat gold with 4 mother-of-pearl

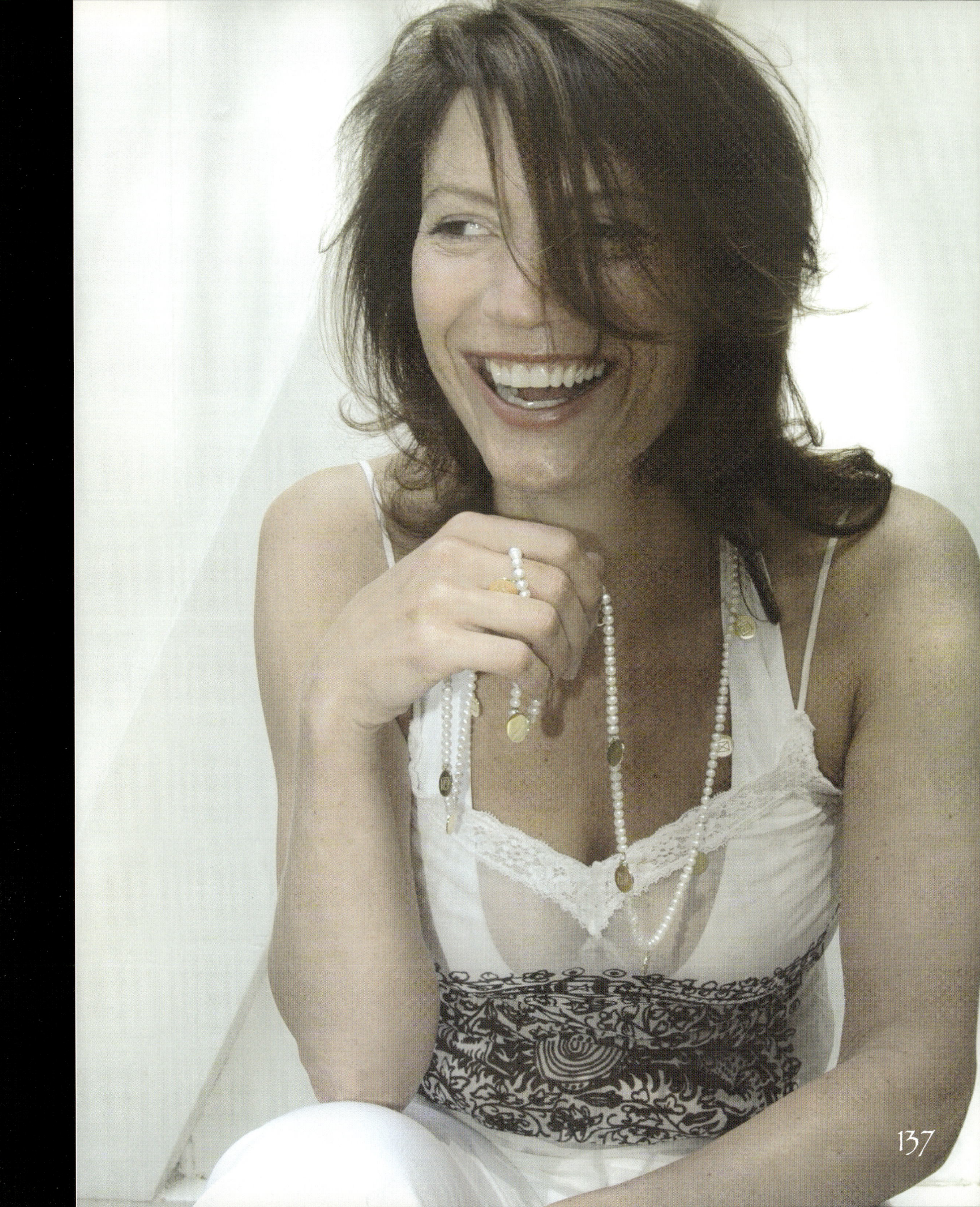

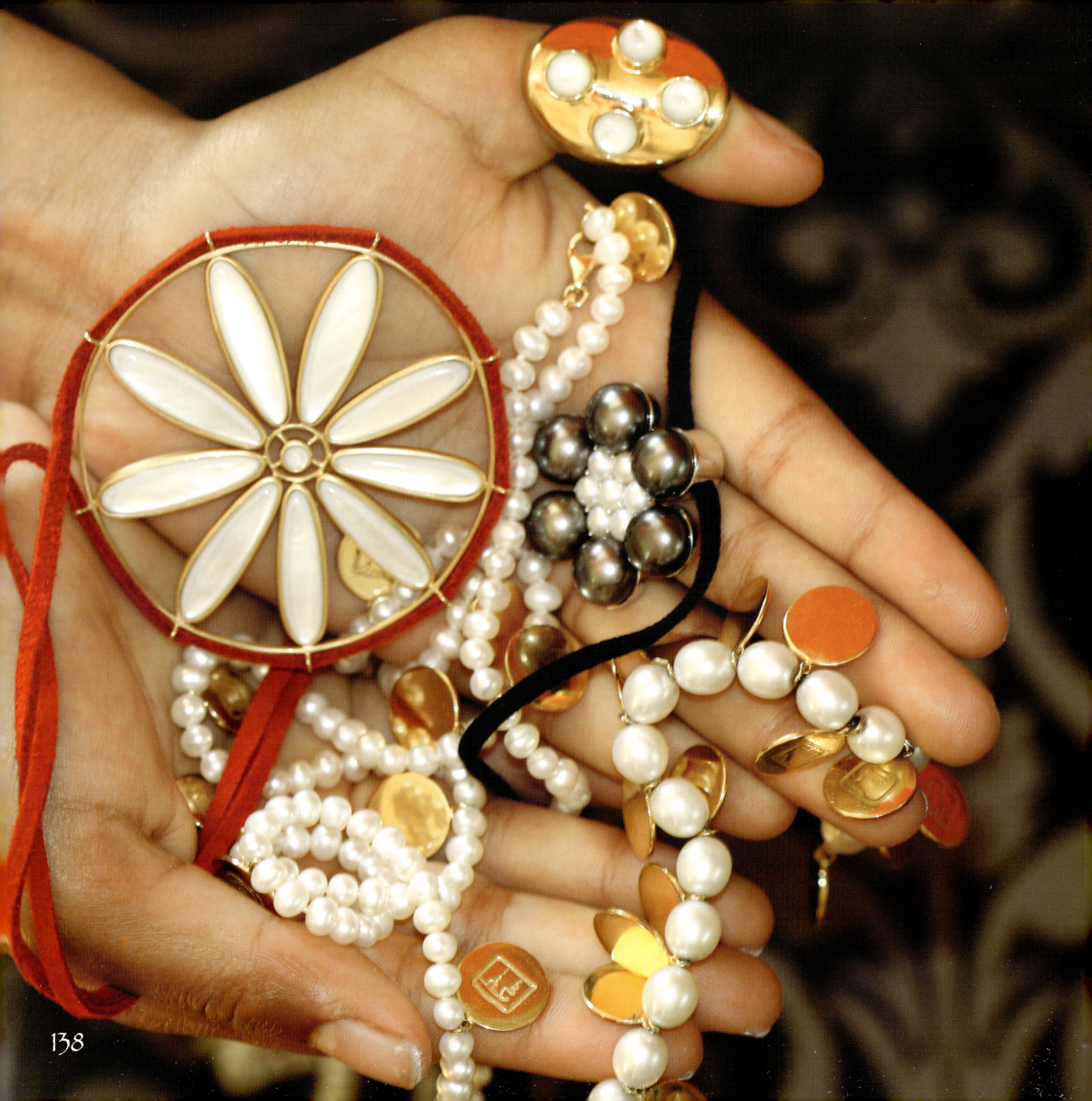

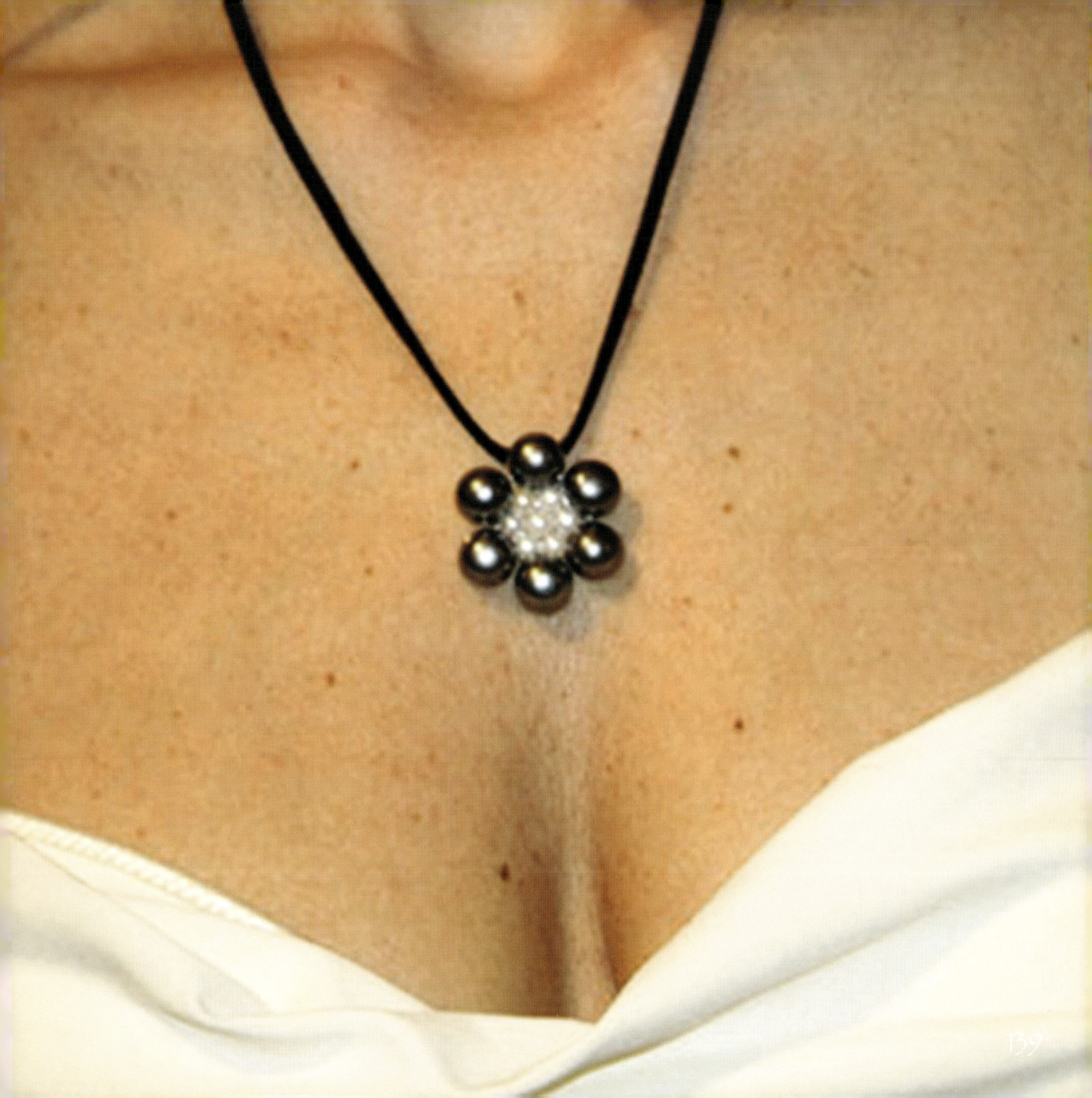

"Tzolkin Figure"

eighteen-carat gold figure with

13 moveable joints

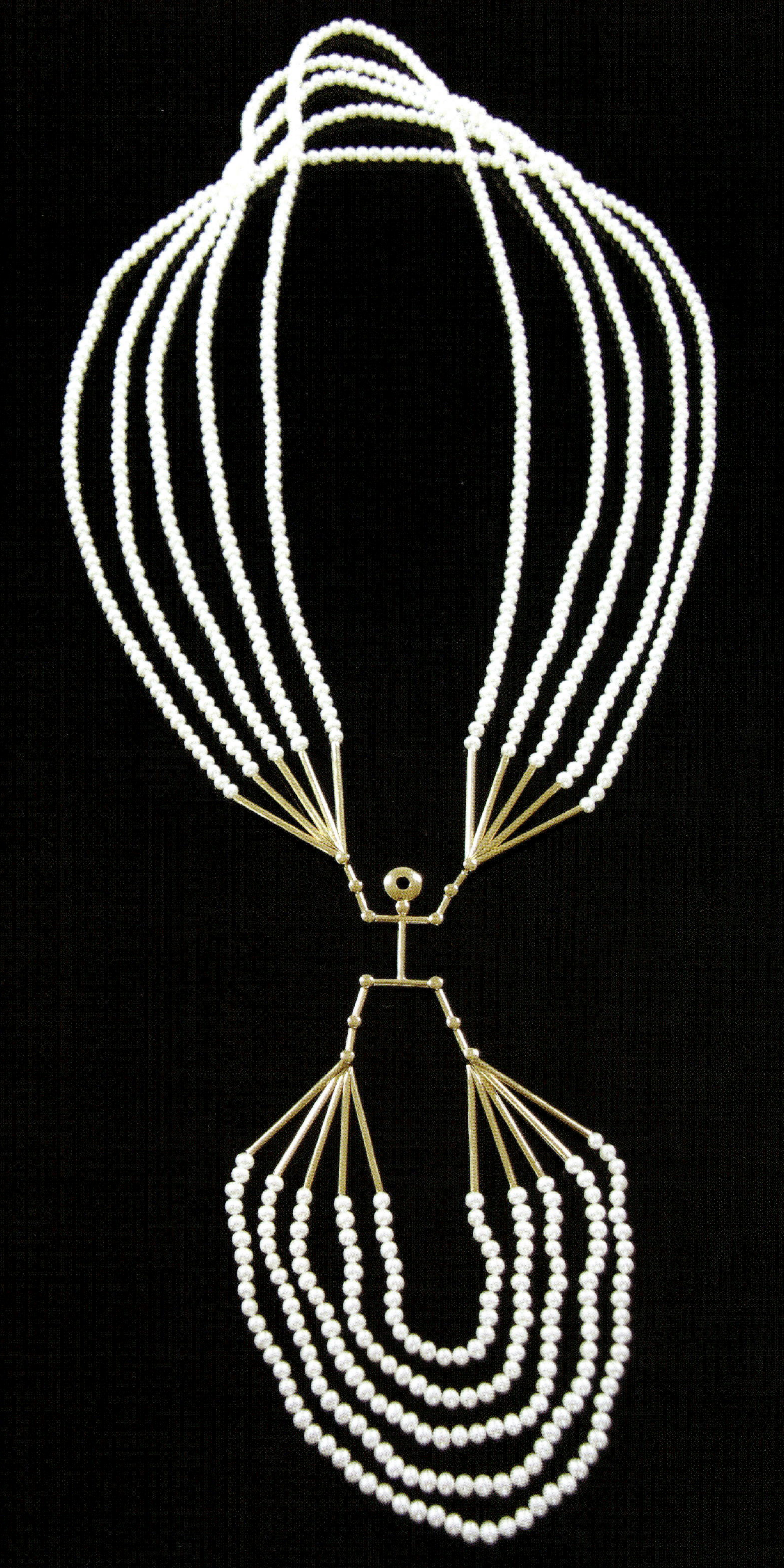

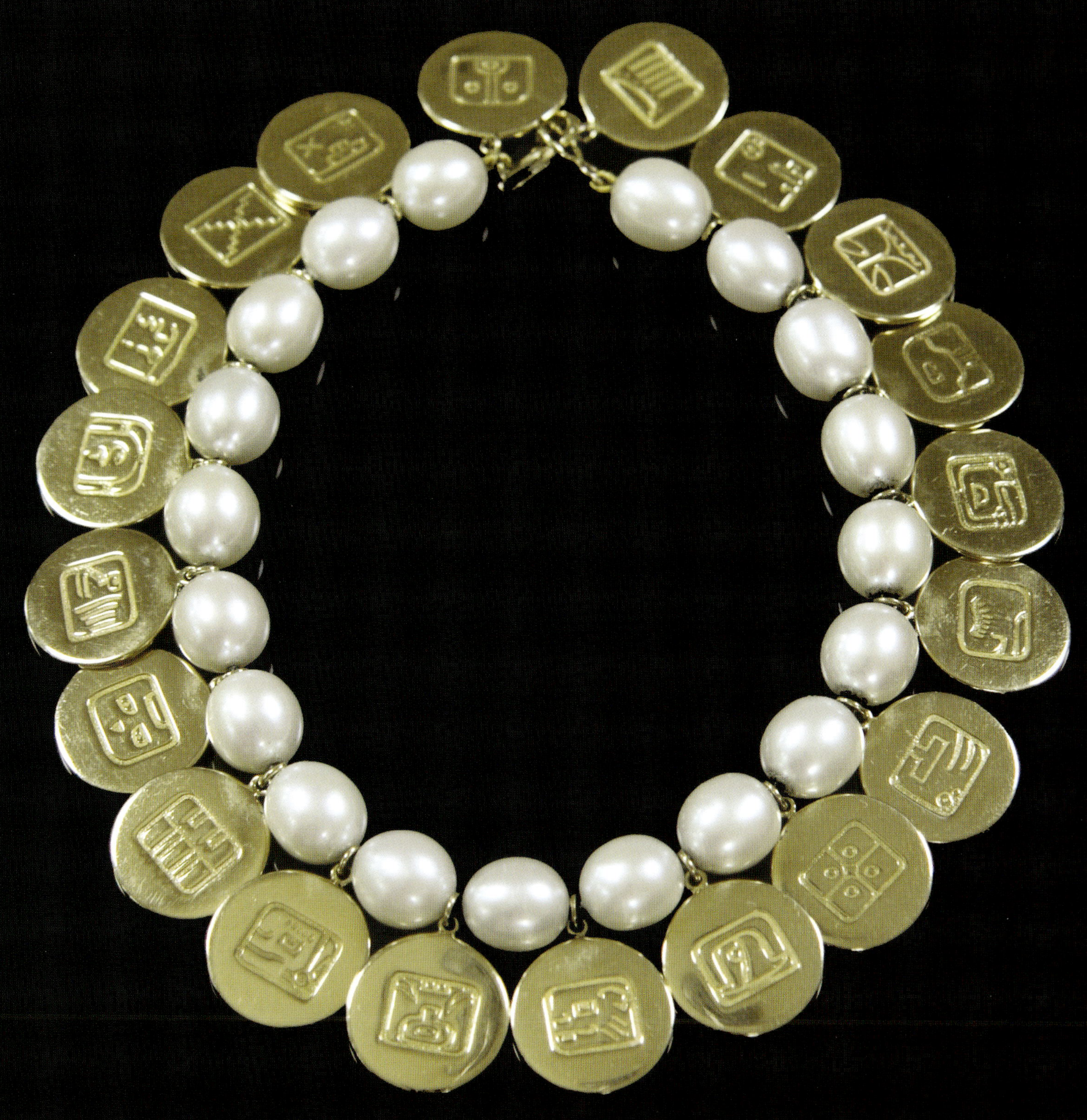

"Uinals Bracelet"

20 eighteen-carat gold coins with pearls

13 moon zodiac calendar with Tzolkin Count

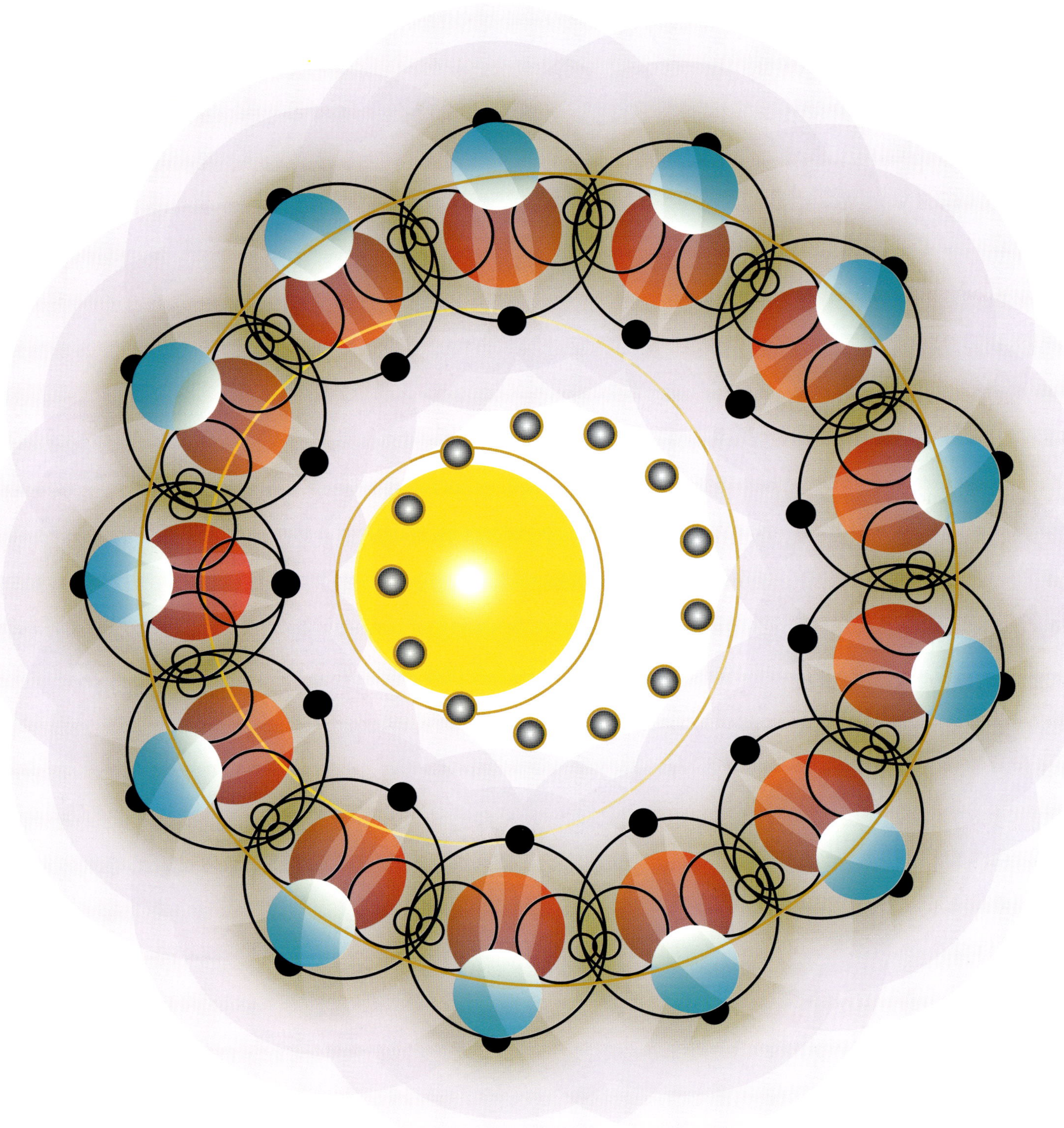

Why change our calendar?

Time has come to make a fundamental change. What we think defines the time we live in. If we change time, will the way we think change too?

If we synchronise our daily lives with our surrounding nature, the connection will become obvious again. More knowledge of the changing natural influences will lead to healthier choices.
A calendar can help to make the relationship between man, his own nature and the surrounding nature clear. There are 13 full moons a year and women get their period 13 times a year.
A calendar of 13 months represents the natural life~cycle on Earth. Many old cultures had moon based calendars of 13 months. Living according to the Moon influences is beneficial, because you do not turn unnecessary against natural forces. Our Gregorian calendar with 12 months, commonly used today, is implemented by Julius Caesar. He used the calendar as a tool to suppress large communities by pulling them away from their natural rhythm. He named the months July and August after himself and attributed these months the highest number of days, taken of February. Pope Gregory also altered the calendar a bit in favor of the Catholic Church in 1582.

We see time as a horizontal line from past to present and merely as a convenient framework for our working lives. This approach has led us to our industrial age, which now threatens the survival of all life on Earth. We need to change our viewpoint but we cannot create a new world without changing our concept of time. Our Earth and solar system are moving rapidly through huge frequency shifts, time, as we have known it is collapsing and the need for change is no longer just an option.

The Ancient Maya were masters of the true science of time and left behind vital keys to unlock this great mystery, with their use of symbols and infinite knowledge of the cosmos.

They carefully recorded the closing of our present 26,000 year great cycle in 2012. No other culture except the Mayan have possessed such an advanced calendar knowledge and for these reasons the calendar in this book is implemented with the Mayan system known as the Tzolkin.

The Mayan Tzolkin Calendar is an infinite study, deeply rooted in the nature of consciousness, enabling us always to go deeper in our understanding. Used as a guide in our daily lives, these harmonics can bring an awareness of the interconnection of all things, from the highest to the lowest levels. Consciousness is not just the prerogative of Man, but also the basis of life in all its forms and of time itself. The information presented here comes as much from the future as from the past, and is sincerely offered to assist people in the industrialised world to rediscover their unity with all life, whilst honoring all perspectives and all cultures.

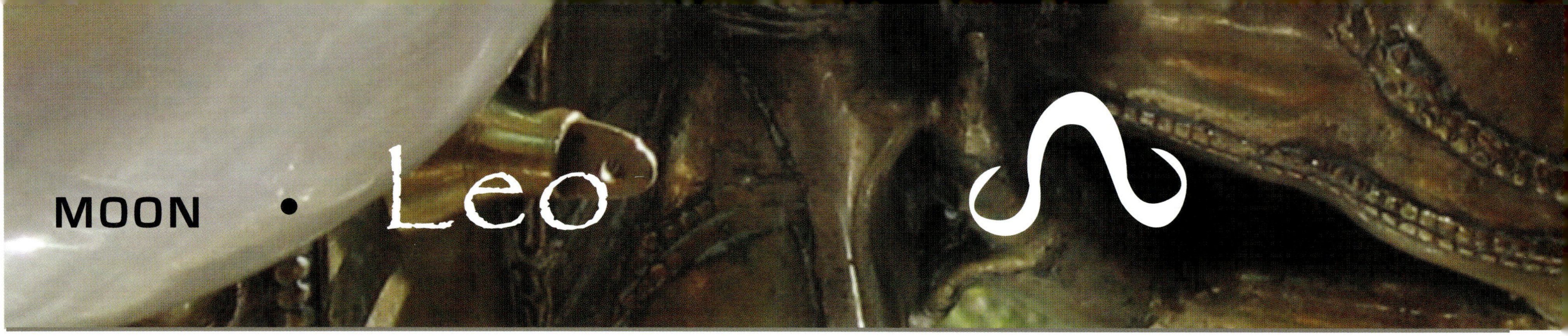

MOON • Leo ♌

Leo 1	Leo 2	Leo 3	Leo 4	Leo 5	Leo 6	Leo 7
2007 July 26 2012	2007 July 27 2012	2007 July 28 2012	2007 July 29 2012	2007 July 30 2012	2007 July 31 2012	2007 August 1 2012
2008 2009 2010 2011	2008 2009 2010 2011	2008 2009 2010 2011	2008 2009 2010 2011	2008 2009 2010 2011	2008 2009 2010 2011	2008 2009 2010 2011

Leo 8	Leo 9	Leo 10	Leo 11	Leo 12	Leo 13	Leo 14
2007 August 2 2012	2007 August 3 2012	2007 August 4 2012	2007 August 5 2012	2007 August 6 2012	2007 August 7 2012	2007 August 8 2012
2008 2009 2010 2011	2008 2009 2010 2011	2008 2009 2010 2011	2008 2009 2010 2011	2008 2009 2010 2011	2008 2009 2010 2011	2008 2009 2010 2011

Leo 15	Leo 16	Leo 17	Leo 18	Leo 19	Leo 20	Leo 21
2007 August 9 2012	2007 August 10 2012	2007 August 11 2012	2007 August 12 2012	2007 August 13 2012	2007 August 14 2012	2007 August 15 2012
2008 2009 2010 2011	2008 2009 2010 2011	2008 2009 2010 2011	2008 2009 2010 2011	2008 2009 2010 2011	2008 2009 2010 2011	2008 2009 2010 2011

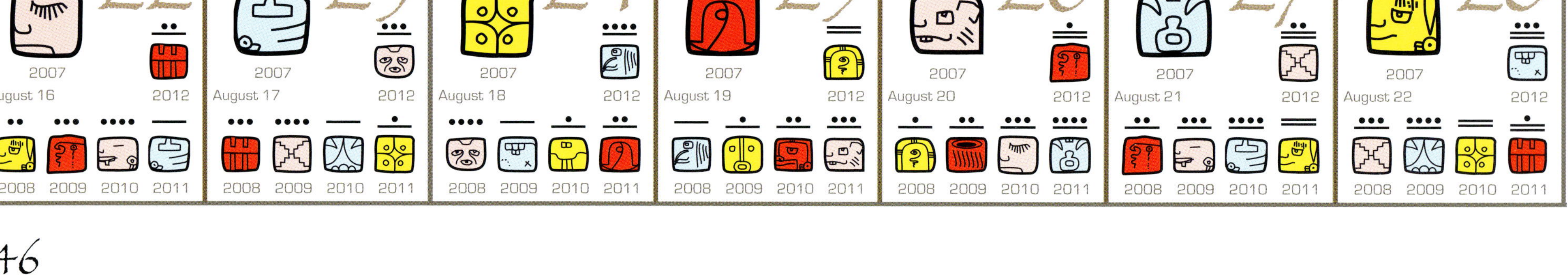

Leo 22	Leo 23	Leo 24	Leo 25	Leo 26	Leo 27	Leo 28
2007 August 16 2012	2007 August 17 2012	2007 August 18 2012	2007 August 19 2012	2007 August 20 2012	2007 August 21 2012	2007 August 22 2012
2008 2009 2010 2011	2008 2009 2010 2011	2008 2009 2010 2011	2008 2009 2010 2011	2008 2009 2010 2011	2008 2009 2010 2011	2008 2009 2010 2011

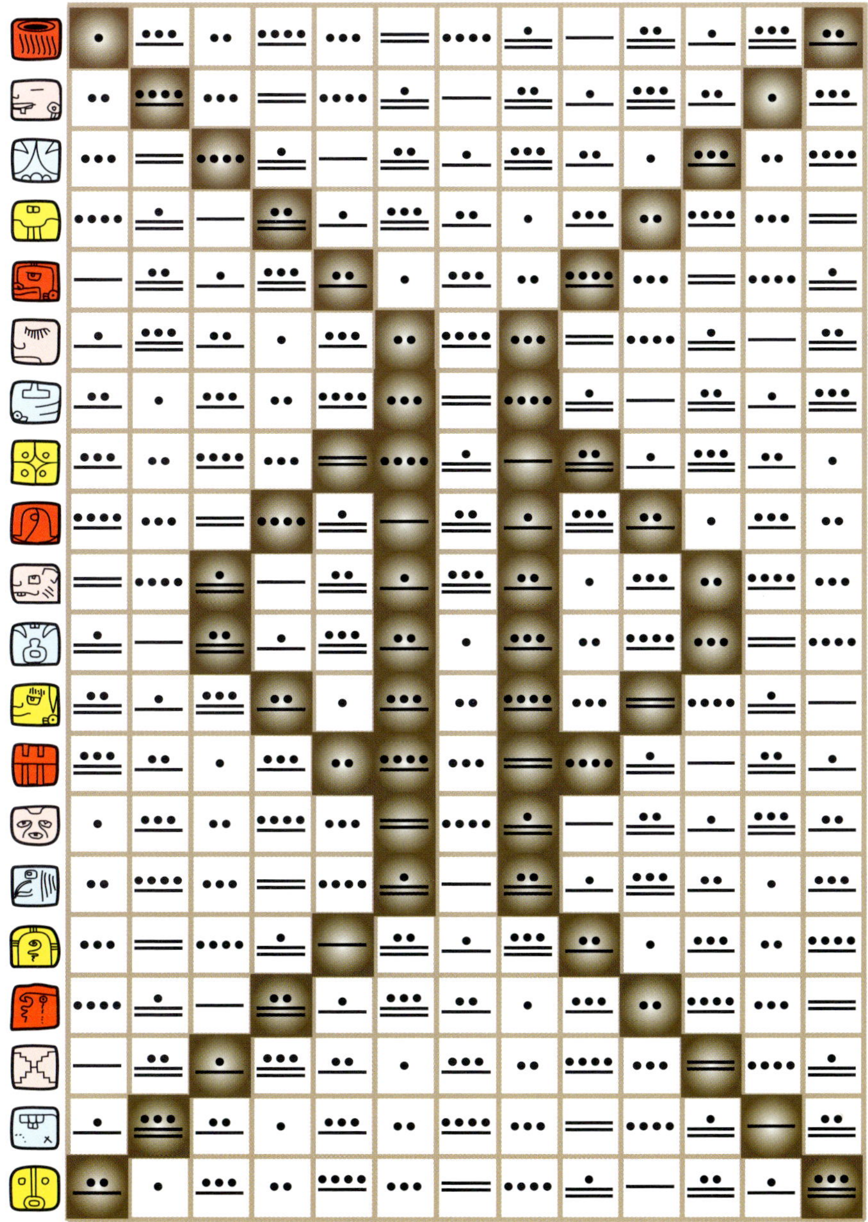

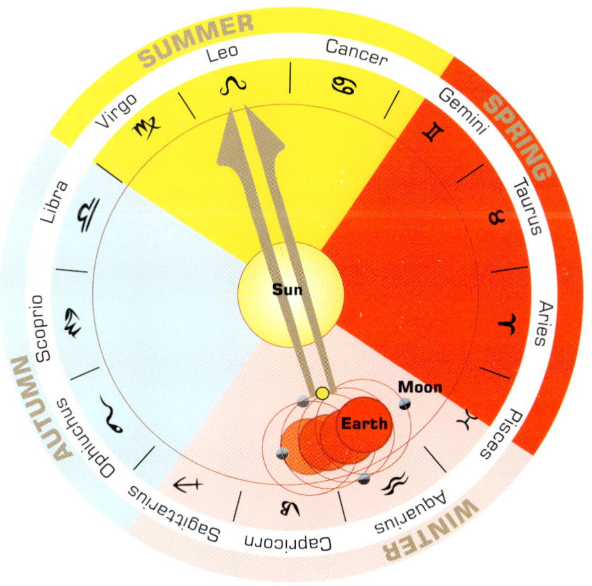

The Tzolkin grid

The Tzolkin grid shows a pattern of thirteen vertical columns representing movement and twenty horizontal rows representing measure. On the left hand side of the first column are 20 sun glyphs, each one representing a unique pulse of the sun. We were all born under the energy of one of these sun symbols and one of the 13 tones of movement. Our birthdays lie in one of those squares. The 52 dark squares are the highly charged portal days giving access to multiple levels of energy. The dots and bars in the squares refer to the Mayan counting system, a dot • = 1 and a bar —— = 5. Beginning at the top left hand corner the Tzolkin waves flow downwards repeating a 1 to 13 cycle, always flowing from the top downwards until finally reaching square 260, which is 13 Ahau and then the cycle repeats itself again. All the sun glyphs have been given their Mayan mantric names to further enhance their energy resonance.

Throughout our lives we have been conditioned to believe in a linear concept of time, rather than the spiral harmony of our galaxy. For the Maya this energy was symbolised by the form of a sacred G. This is frequently found in their artwork symbolising our place of origin in the Milky Way, the Hunab Ku, the source of all life.

It is important to become aware now of embracing a 13-day spiral wave in our life, the essence of the universe. The experience is somewhat like surfing; first we need to know where to catch the wave. This can be easily derived from the calendar by remembering that each new wave begins with tone one (one dot over the symbol) and that this symbol's energy governs the whole wave.

Tones one to five are the planning and preparation period. Tones six to eleven are the action and completion stages. Twelve is for viewing and future planning. Thirteen brings rest, renewal and recharge for the next wave spiral. When we plan our daily lives and goals riding the wave thirteen, we embrace the natural rhythm and synchronicity of the Universe. The merging of our consciousness with these Mayan symbols and time waves open up profound new levels of awareness that many people have already experienced over the past decade.

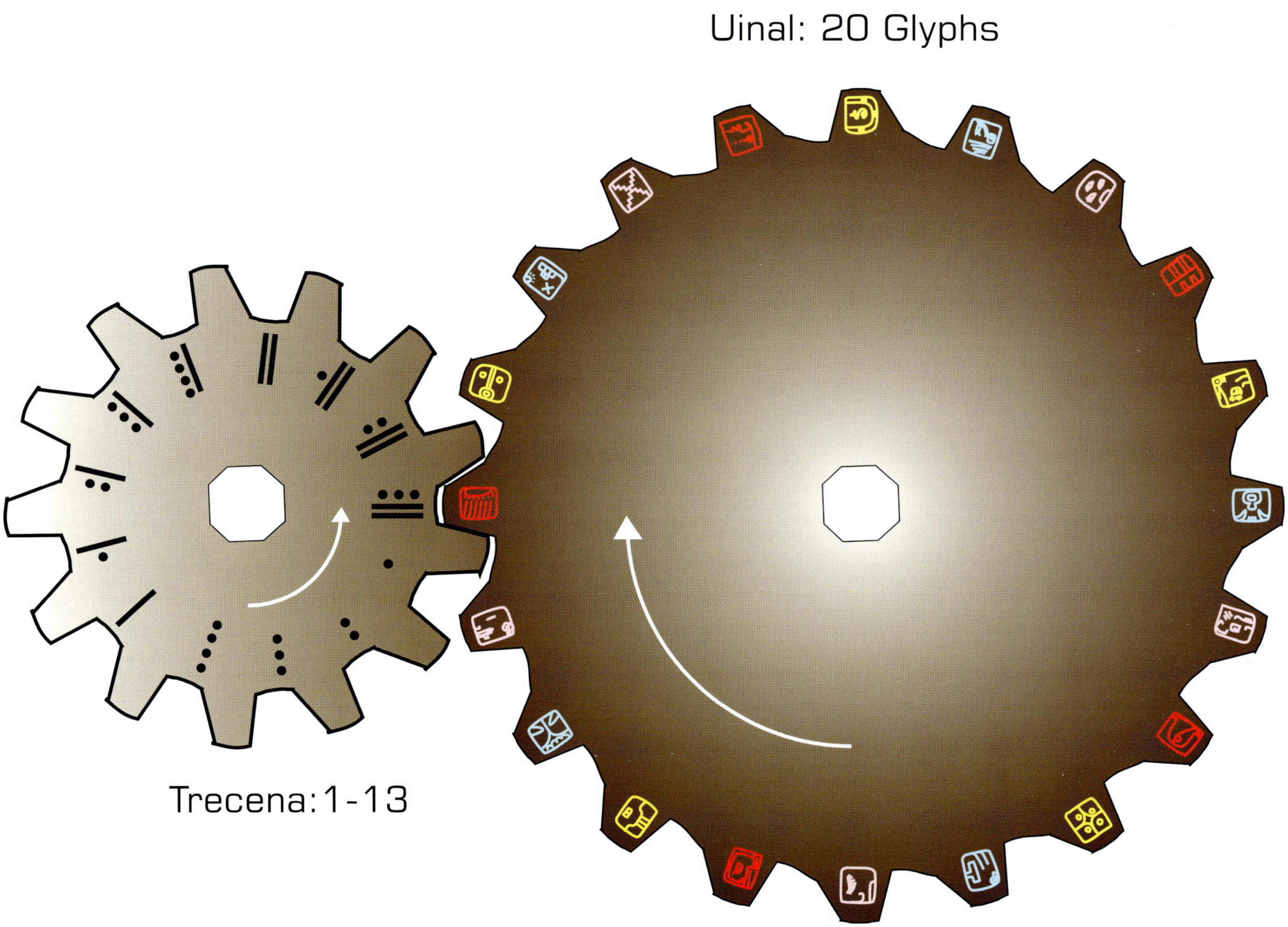

Uinal: 20 Glyphs

Trecena:1-13

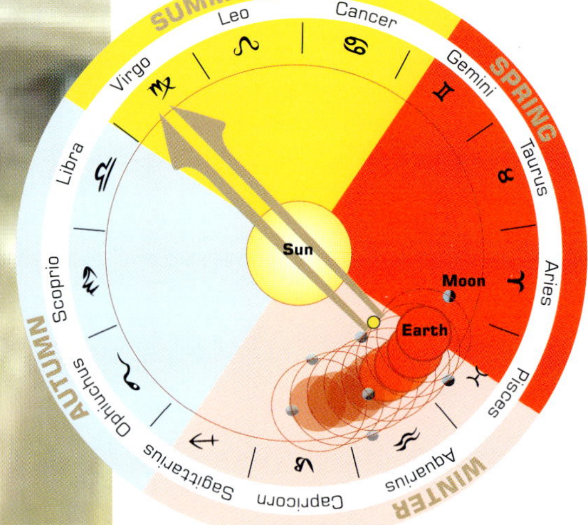

The Tzolkin Wheels

The Tzolkin is the most sacred of the Maya calendars. A count of 260 days, based on the trecena, the numbers 1 to 13; after 13 it starts again with 1. The Uinal consists of 20 glyphs (signs) also in a specific order.

Each day is assigned a number and a glyph. A Tzolkin round consists of 13 x 20 = 260 days. The Quiché-Maya have kept this calendar intact for 2,500 years through the diligent observance of the calendar by so-called day-keepers, men and women endowed with the special responsibility of keeping track of the days. Through comparisons with dates on old sites, archaeologists have been able to verify that not one single day has been lost in 2,500 years. In this book the Tzolkin count is operating from 2007 until December 21 2012.

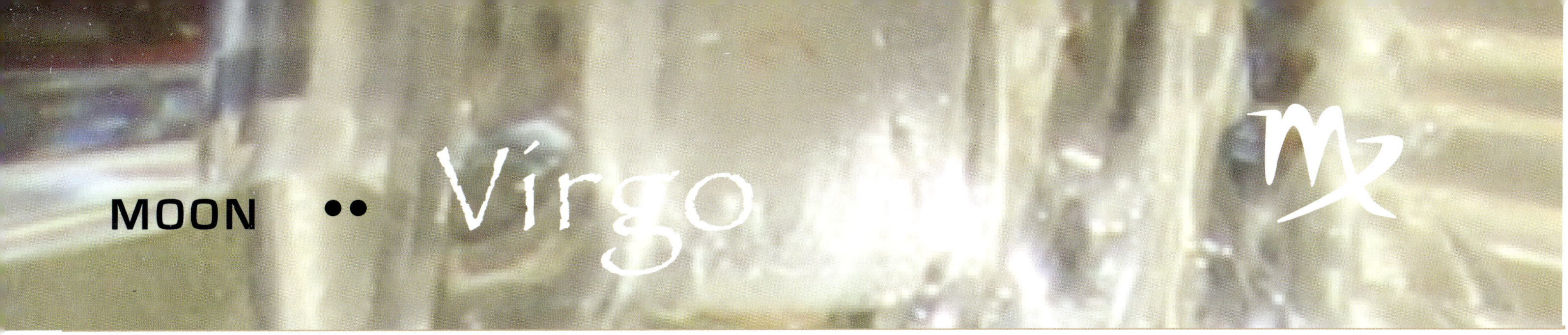

MOON •• Virgo ♍

Virgo	2007	2012	2008	2009	2010	2011
1	august 23	2012	2008	2009	2010	2011
2	august 24	2012	2008	2009	2010	2011
3	august 25	2012	2008	2009	2010	2011
4	august 26	2012	2008	2009	2010	2011
5	august 27	2012	2008	2009	2010	2011
6	august 28	2012	2008	2009	2010	2011
7	august 29	2012	2008	2009	2010	2011
8	august 30	2012	2008	2009	2010	2011
9	august 31	2012	2008	2009	2010	2011
10	september 1	2012	2008	2009	2010	2011
11	september 2	2012	2008	2009	2010	2011
12	september 3	2012	2008	2009	2010	2011
13	september 4	2012	2008	2009	2010	2011
14	september 5	2012	2008	2009	2010	2011
15	september 6	2012	2008	2009	2010	2011
16	september 7	2012	2008	2009	2010	2011
17	september 8	2012	2008	2009	2010	2011
18	september 9	2012	2008	2009	2010	2011
19	september 10	2012	2008	2009	2010	2011
20	september 11	2012	2008	2009	2010	2011
21	september 12	2012	2008	2009	2010	2011

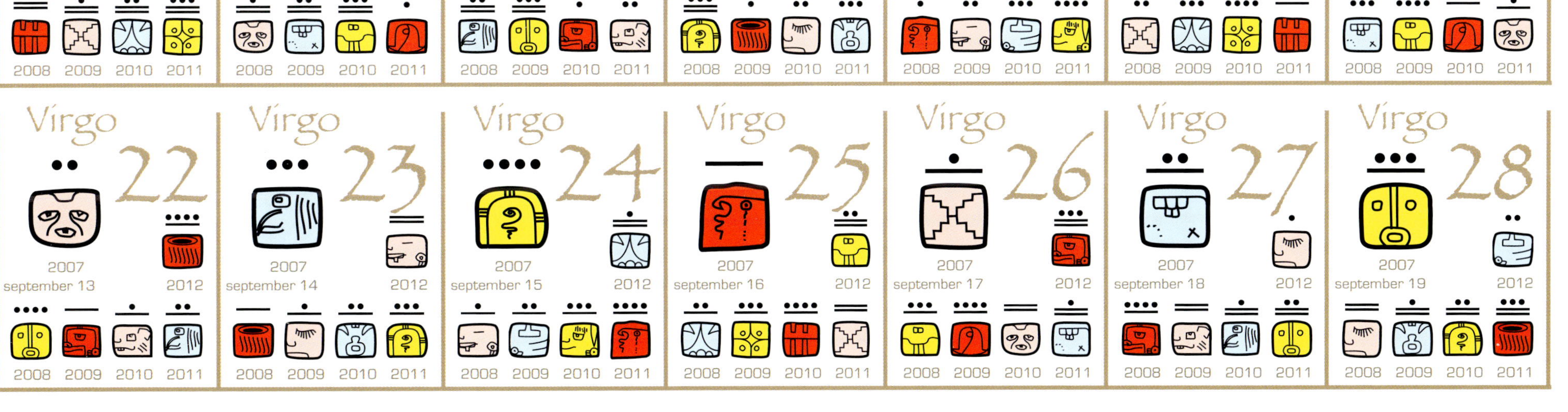

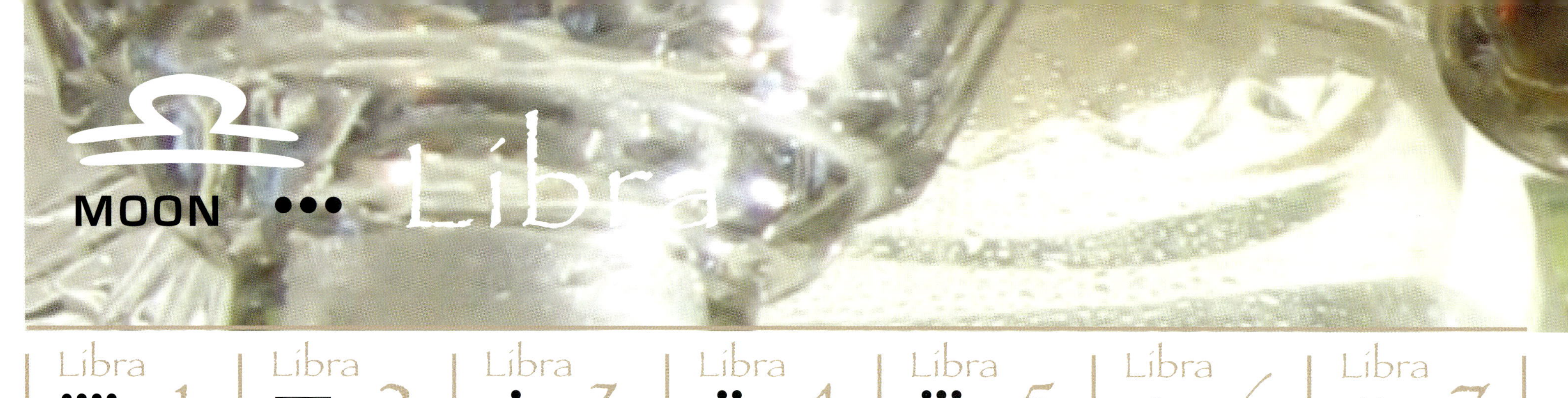

Libra 1	Libra 2	Libra 3	Libra 4	Libra 5	Libra 6	Libra 7
2007 September 20	2007 September 21	2007 September 22	2007 September 23	2007 September 24	2007 September 25	2007 September 26
2012	2012	2012	2012	2012	2012	2012
2008 2009 2010 2011	2008 2009 2010 2011	2008 2009 2010 2011	2008 2009 2010 2011	2008 2009 2010 2011	2008 2009 2010 2011	2008 2009 2010 2011

Libra 8	Libra 9	Libra 10	Libra 11	Libra 12	Libra 13	Libra 14
2007 September 27	2007 September 28	2007 September 29	2007 September 30	2007 October 1	2007 October 2	2007 October 3
2012	2012	2012	2012	2012	2012	2012
2008 2009 2010 2011	2008 2009 2010 2011	2008 2009 2010 2011	2008 2009 2010 2011	2008 2009 2010 2011	2008 2009 2010 2011	2008 2009 2010 2011

Libra 15	Libra 16	Libra 17	Libra 18	Libra 19	Libra 20	Libra 21
2007 October 4	2007 October 5	2007 October 6	2007 October 7	2007 October 8	2007 October 9	2007 October 10
2012	2012	2012	2012	2012	2012	2012
2008 2009 2010 2011	2008 2009 2010 2011	2008 2009 2010 2011	2008 2009 2010 2011	2008 2009 2010 2011	2008 2009 2010 2011	2008 2009 2010 2011

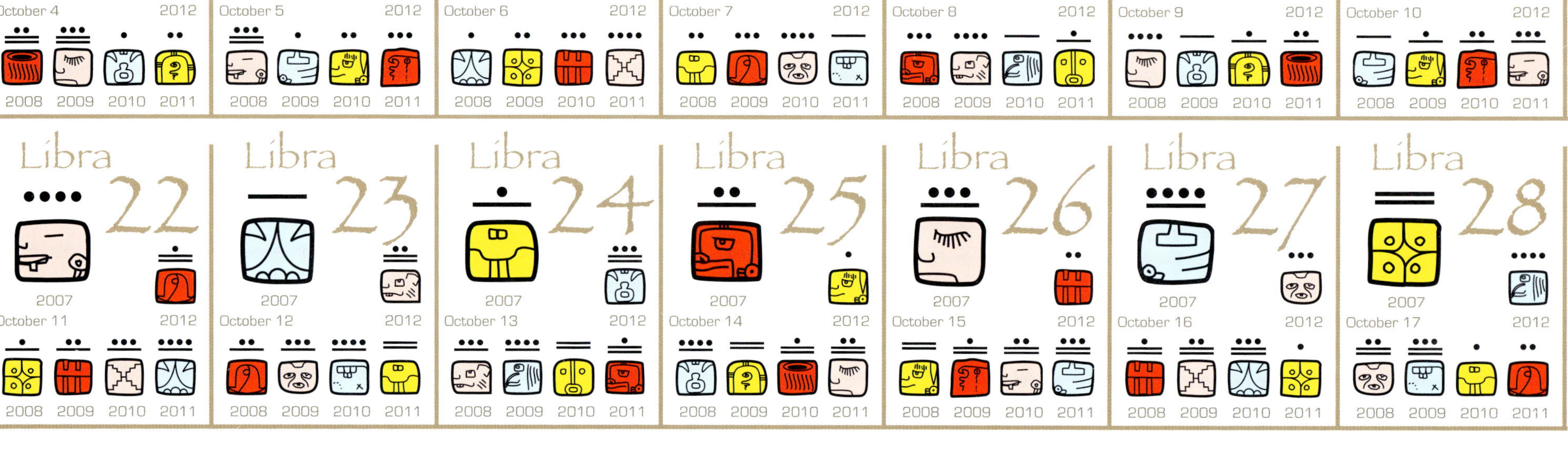

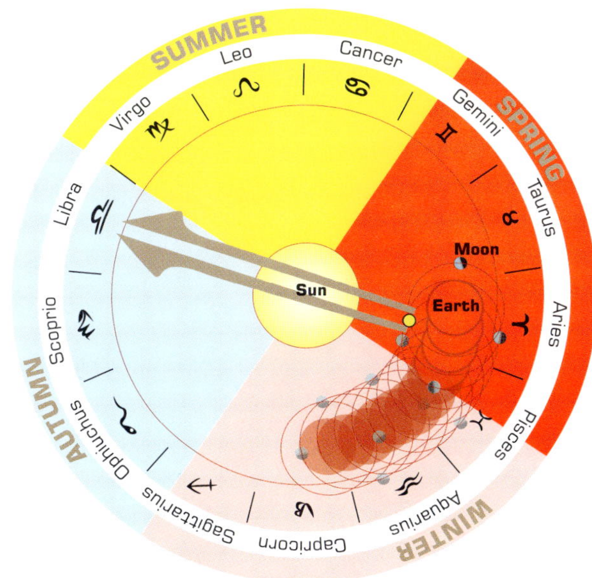

Trecena

Hun: 1. Unity

Ca: 2. Polarity

Ox: 3. Rhythm

Can: 4. Measure

Ho: 5. Center

Uac: 6. Organic Balance

Uc: 7. Mystical Power

Vaxac: 8. Harmonic Resonance

Bolon: 9. Greater Cycles

Lahun: 10. Manifestation

Hun Lahun: 11. Dissonance

Ca Lahun: 12. Complex Stability

Ox Lahun: 13. Universal Movement

Red = EAST, element: water

Imix

Imix is variously translated as 'dragon' or 'crocodile' and is associated with the very beginning of things and the primal unconscious: initiates, births. It is the abundance of potential and all that is not extant. The source of all and the place to which all returns, Imix is the beginning and the end of the transformative cycle. Imix is where energy resides as instinct, memory and potential. planet ~ Neptune. Affirmation: "I trust unconditionally the source of divine nurturance."

Chicchan

Chicchan means 'serpent'. Chicchan is the energy which lies coiled at the base of the spine (kundalini) and calls to our most instinctive drives. Serpent wisdom worldwide is essential amoral, beyond opposites, and is strongly connected with both spiritual and sexual activity. Chicchan is a survivor. planet ~ asteroid belt. Affirmation: "I joyfully access and experience the wisdom and vitality held in my body."

Muluc

Muluc means 'water' and 'that-which-is-gathering-up'. Muluc purifies. It is the universal movement, the cosmic tide. This power can nurture or devastate and has all the preciousness and collective influence of water. This constant movement like water is central to the transformative cycles of all life forms on earth. Muluc is the tide which gathers the individual towards community. planet ~ Mercury Affirmation: "As a galactic beacon and receiver, I have direct access to the Divine."

Ben

Ben is 'pillars of light', 'sky walker' explores space. Ben represents a complex image combining inner- and outer-authority, spiritual and physical motivation. Ben is wakefulness. It is the essence of both conquest and relationship and is the tribal manifestation of primal consciousness. planet ~ Mars
Affirmation: "I am a pillar of the invisible temple that brings heaven to Earth."

Caban

Caban means 'earthquake' and is number 17, associated with mother figure and inspiration - the initiator and source of all creative activity. Caban is a navigator. Caban is the consciousness of change within communities and often provides the ideas which are the precursors of change. planet ~ Uranus. Affirmation: "I am aligned with my center, Earth force aligned with cosmic force."

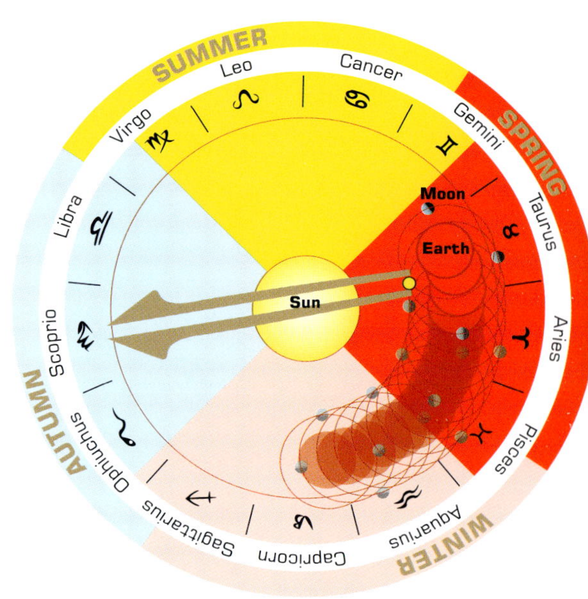

MOON •••• Scorpio

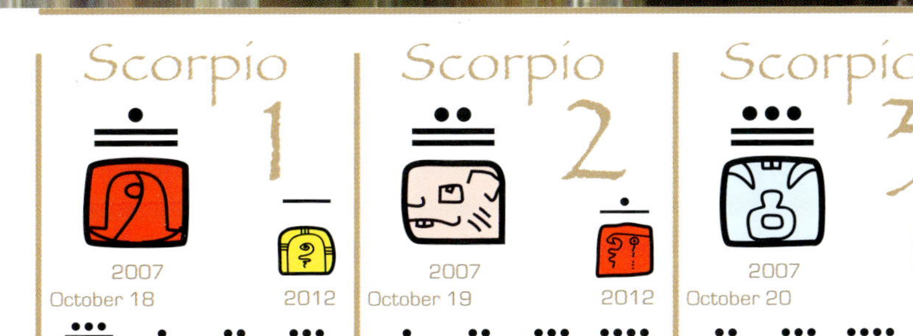

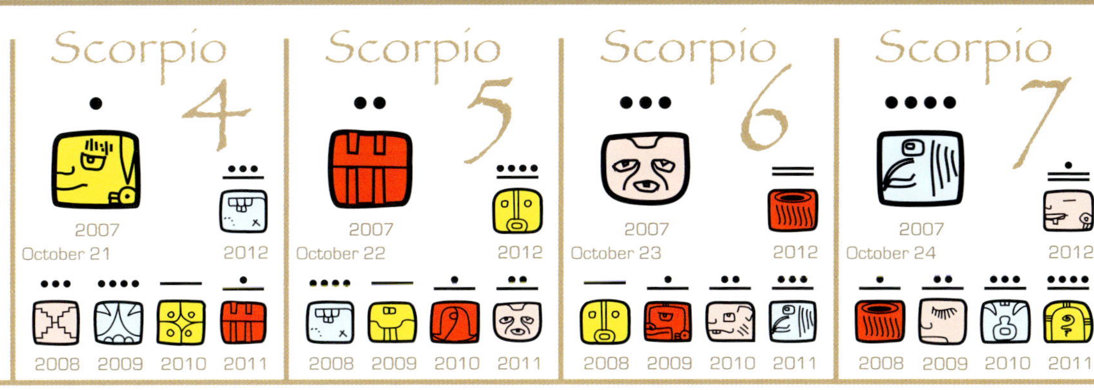

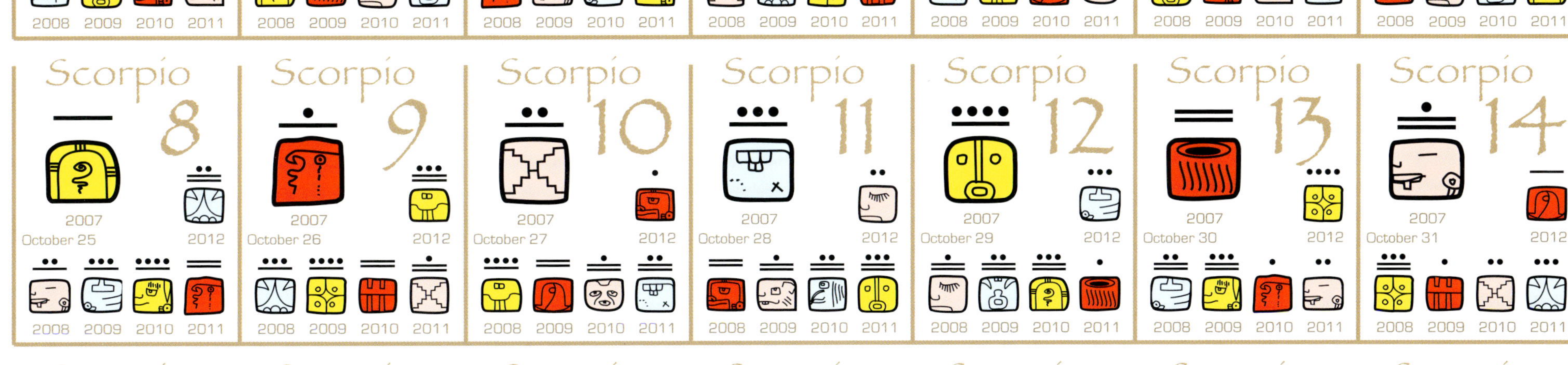

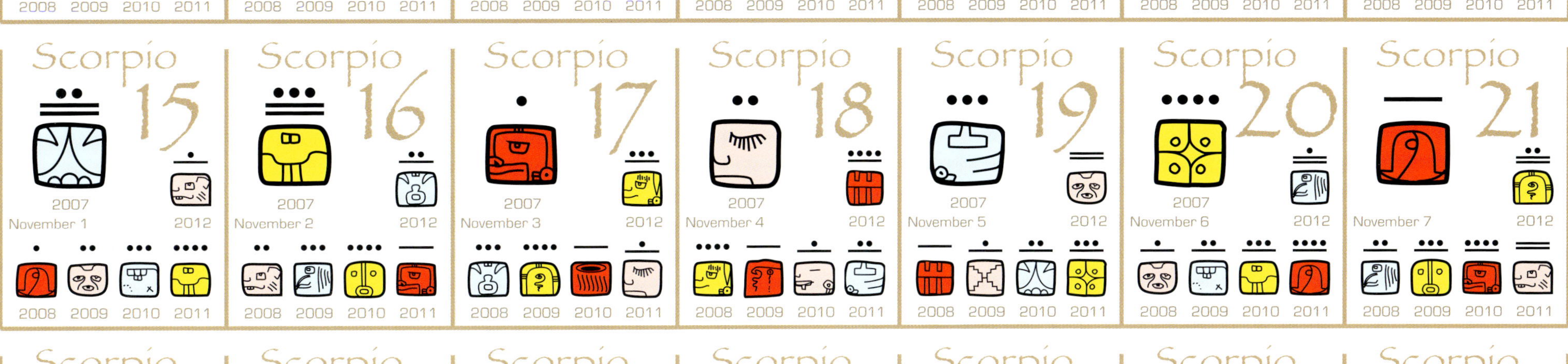

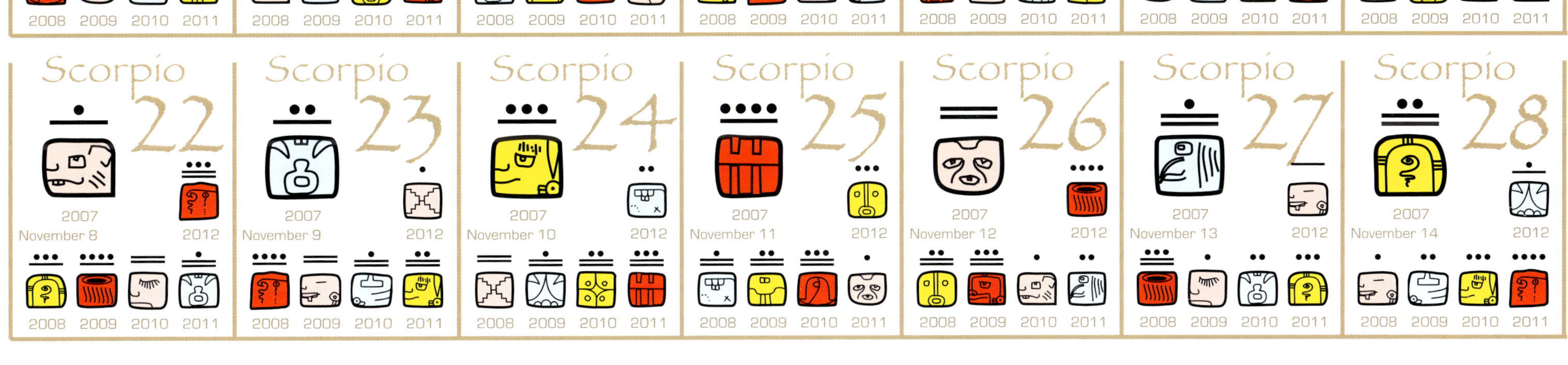

MOON — Ophiuchus

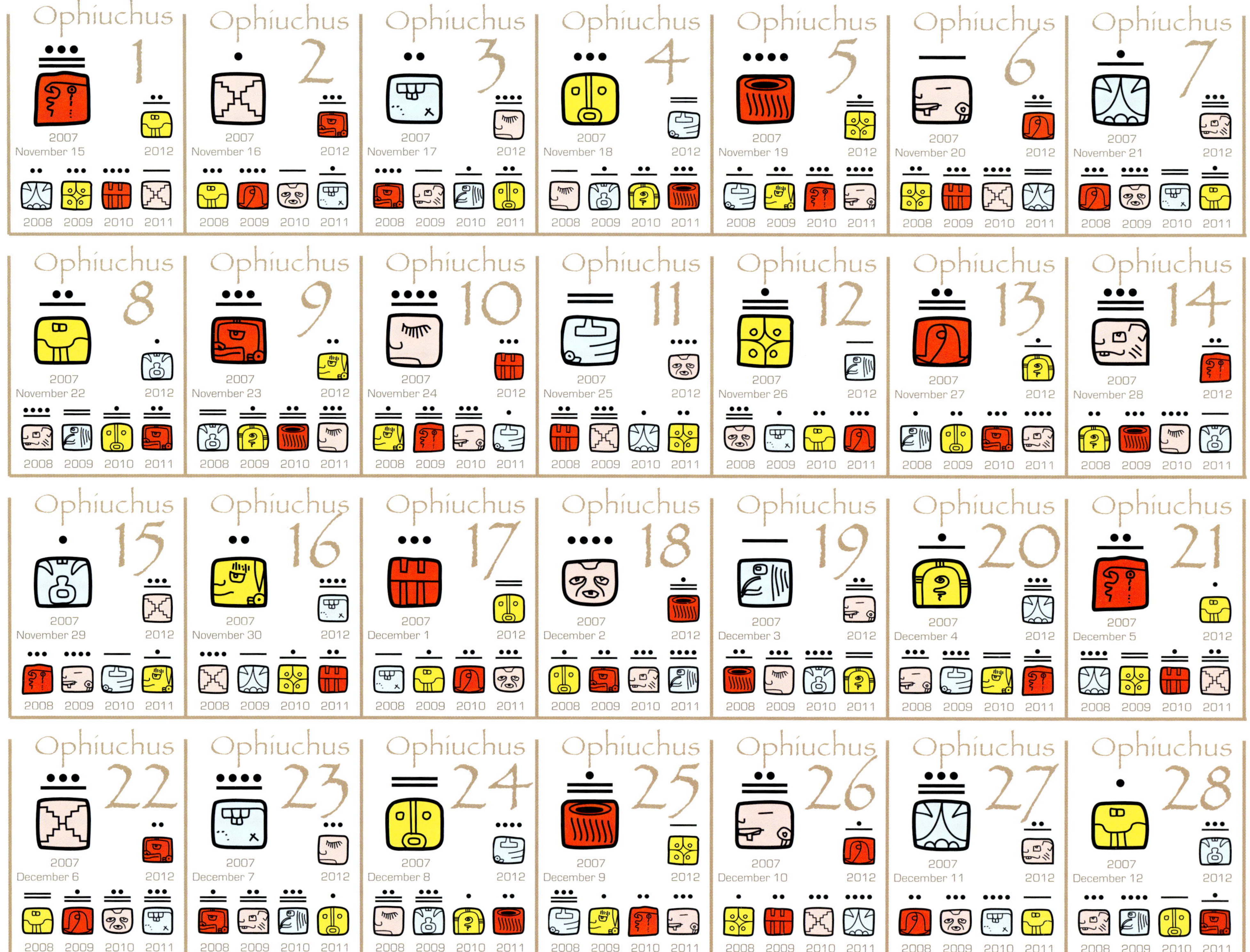

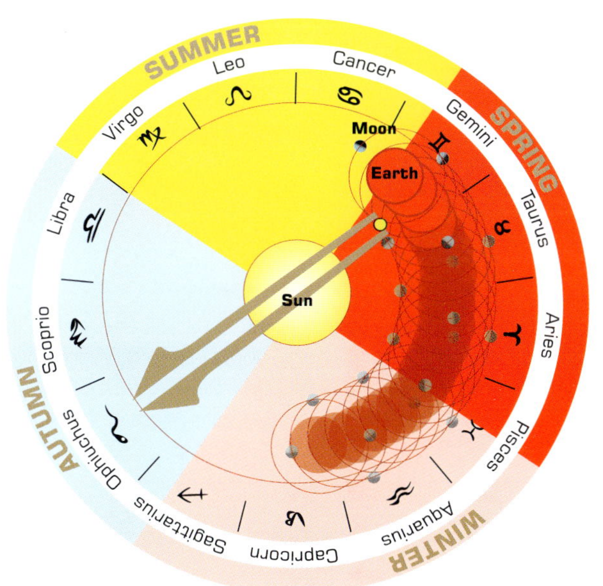

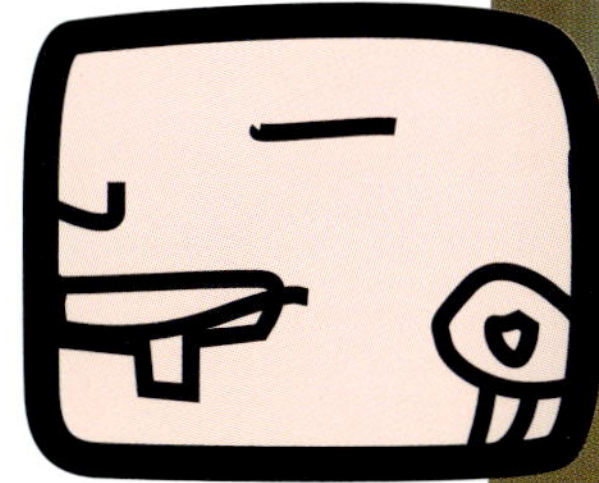

Ik means 'wind' or 'breath'. This second face of the Ahau kin may be imagined as an echo of that primal wind of release which brings vitality in rain and destruction by storm. Ik is the breath which animates us all and the breath through which we communicate. planet ~ Uranus. Affirmation: "I am the unified presence of Spirit."

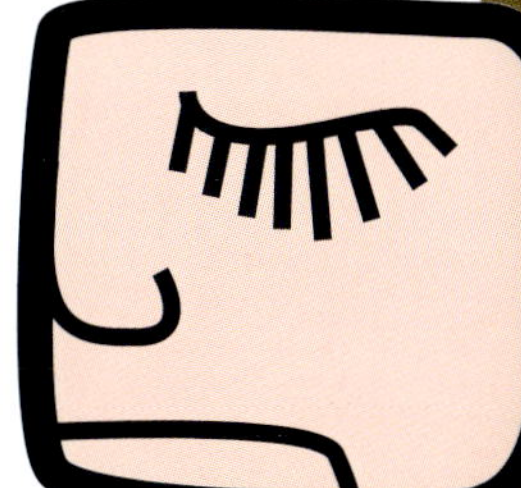

Cimi means 'world bridger' or 'death' and is representative in the act of transformation - both of self and others. Cimi is a bridge between worlds, a listener and communicator - a representative of sacrifice/forgiveness and release. Cimi equalises new opportunities. planet ~Mars. Affirmation: "I release. I let go. I surrender. I forgive."

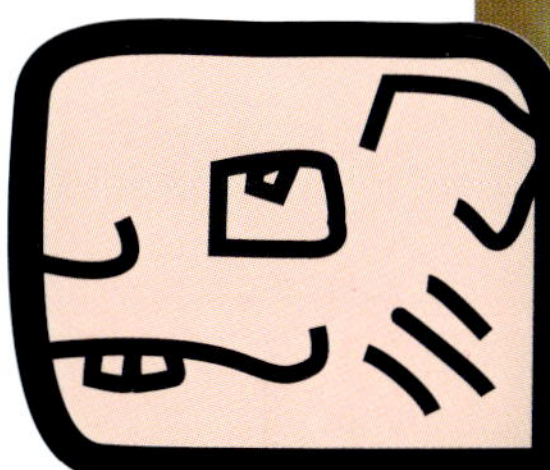

Oc means 'dog' and 'gateway'. Oc is in many ways the archetype of dog: living close to its instincts, loyalty and unconditional love. Oc is our guide through the transformative darkness. Oc is the power which navigates our private underworld, where logic and intellect fail. In Maya myths Oc's footprints mark the very beginning and end of time. planet ~ Mercury. Affirmation: "I open to new beginnings. I receive guardians and guides."

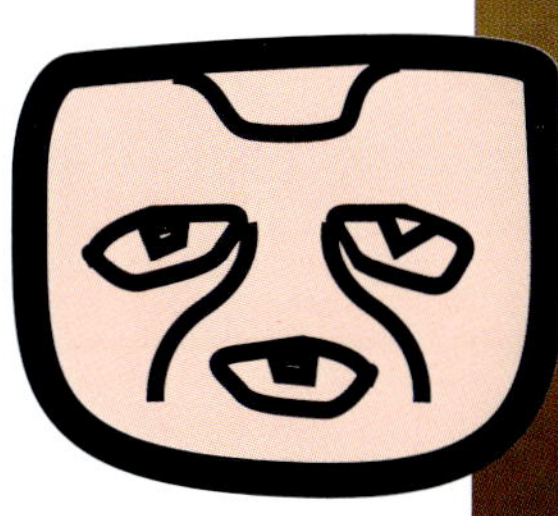

Ix means 'jaguar' or 'wizard'. Ix enchants. Ix is the temporal and spiritual wealth which brings consciousness to the world and is associated with the night and our intuitive selves. Ix is a powerful hunter, lord of our transmutation. Ix is sacred energy which underlines earthquakes and storms. Under the shamanic and transformative influence of Ix, we remember our cosmic identity again. planet ~ Asteroid belt. Affirmation: "Through heart-knowing, I am in natural alignment with divine will."

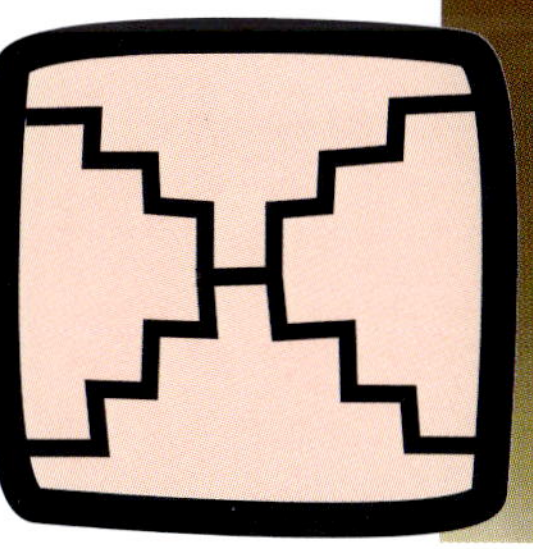

Etznab means 'mirror' and 'knife', a sword of truth - the very edge of the polarity: light and darkness. It is the awareness of opposites and thus is the essence of resolution. Etznab shares characteristics both with separation and the mirror of reflection and reflected. This is the possibility of reunification - depolarisation - 'paradise regained' - embodied in the evolutionary times and souls of earthkind. planet ~ Pluto. Affirmation: "I am the truth beyond mirrors. I now step through the mirror into the greater reality."

Blue = West; element earth

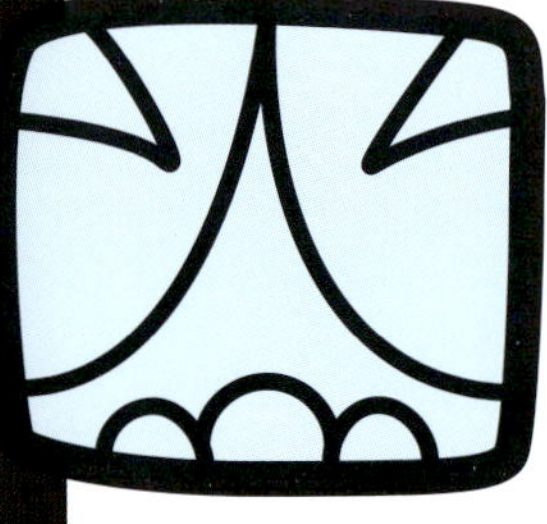

Akbal means 'night' and 'darkness'. Akbal is abundance. It is redolent of dark places under the Earth, the lightless temples atop the pyramids of ancient Maya and of the dark spaces within our own minds. All these are places of great spiritual and physical wealth. planet ~ Saturn. Affirmation: "I pilgrimage deep within, to the sanctuary of self, to garner the gifts awaiting me there."

Manik means 'hand' or 'deer'. The glyph shows the hand of the hunter. Manik is the powerful and transformative 7th kin, associated with accomplishment and craftsmanship in life. A successful hunt requires that the hunter fully understands - becomes one with - the life energy of the prey. This balance between hunter and hunted is imagined as a wholeness. planet ~ Earth. Affirmation: " I complete. I open. I am one with the light."

Chuen means 'monkey' and is associated with legends about artificers and craftsmen. This kin embodies all the arts and creative excellence. Chuen links sacred knowledge with creative artistry. Chuen is an intuitive trickster, humourist and performer. Chuen days are charged with innate knowledge of relationships. planet ~ Venus. Affirmation: " I am an innocent, open-hearted, transparent, divine child."

Men means 'eagle' - a totem of shamans worldwide - or 'knower'. This is one of the strongest faces of the 20 kin. Men is associated with truth, far-sight and the ability to soar beyond confinement. Men is a sign of strong spirit powerfully motivated. Akbal, Cimi, Ix are transformers from intuition/darkness, Men brings transformed consciousness in clear light. Men comes with the vision and purpose of a spiritual warrior and the sharp discernment which comes from unique perspective. planet ~ Jupiter. Affirmation: "With vision and hope, I dance and sing here for the one heart!"

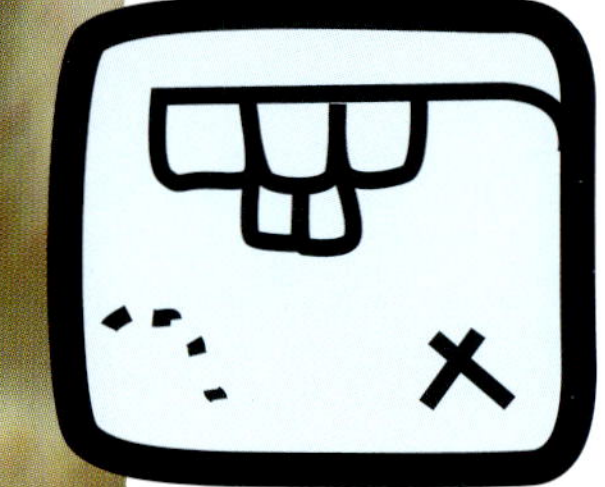

Cauac means 'storm'. Cauac is a self generator. Cauac collects intensifies and releases waves of change - it shares the watery relentlessness of Muluc with the strategy and inspiration of Caban. From the heart of the mountain to the top of the pyramid of heaven, Cauac at once encourages and embodies the becoming of that which is to be. planet ~ Pluto. Affirmation: "I am purifying and transforming myself, igniting my light body, healing all separation."

MOON Sagittarius

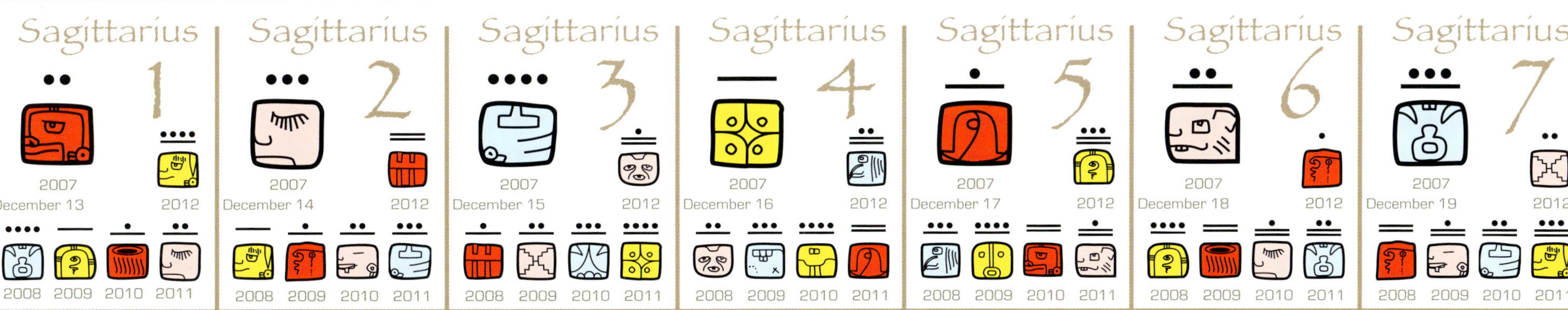

Sagittarius	Year	Date	2012	Years
1	2007	December 13	2012	2008 2009 2010 2011
2	2007	December 14	2012	2008 2009 2010 2011
3	2007	December 15	2012	2008 2009 2010 2011
4	2007	December 16	2012	2008 2009 2010 2011
5	2007	December 17	2012	2008 2009 2010 2011
6	2007	December 18	2012	2008 2009 2010 2011
7	2007	December 19	2012	2008 2009 2010 2011

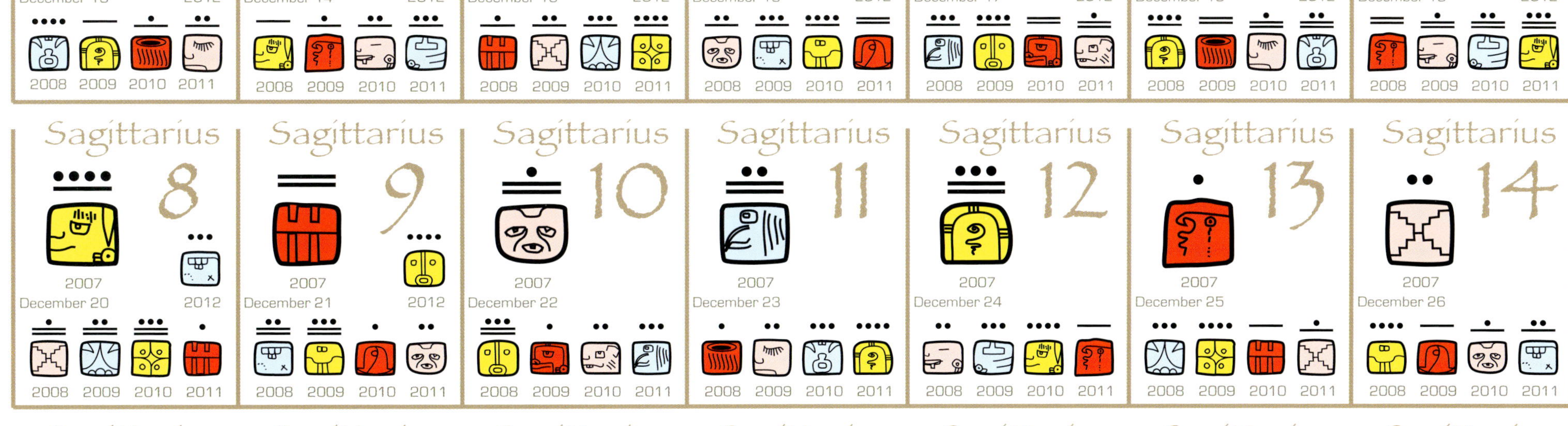

Sagittarius	Year	Date	2012	Years
8	2007	December 20	2012	2008 2009 2010 2011
9	2007	December 21	2012	2008 2009 2010 2011
10	2007	December 22		2008 2009 2010 2011
11	2007	December 23		2008 2009 2010 2011
12	2007	December 24		2008 2009 2010 2011
13	2007	December 25		2008 2009 2010 2011
14	2007	December 26		2008 2009 2010 2011

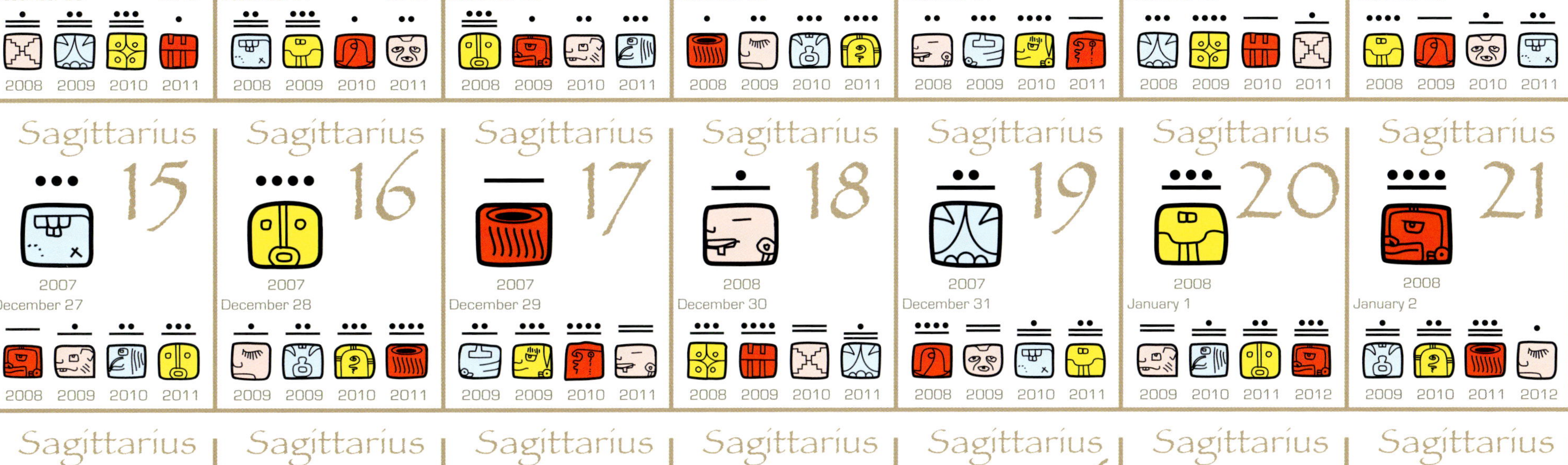

Sagittarius	Year	Date	Years
15	2007	December 27	2008 2009 2010 2011
16	2007	December 28	2009 2009 2010 2011
17	2007	December 29	2009 2009 2010 2011
18	2008	December 30	2008 2009 2010 2011
19	2007	December 31	2008 2009 2010 2011
20	2008	January 1	2009 2010 2011 2012
21	2008	January 2	2009 2010 2011 2012

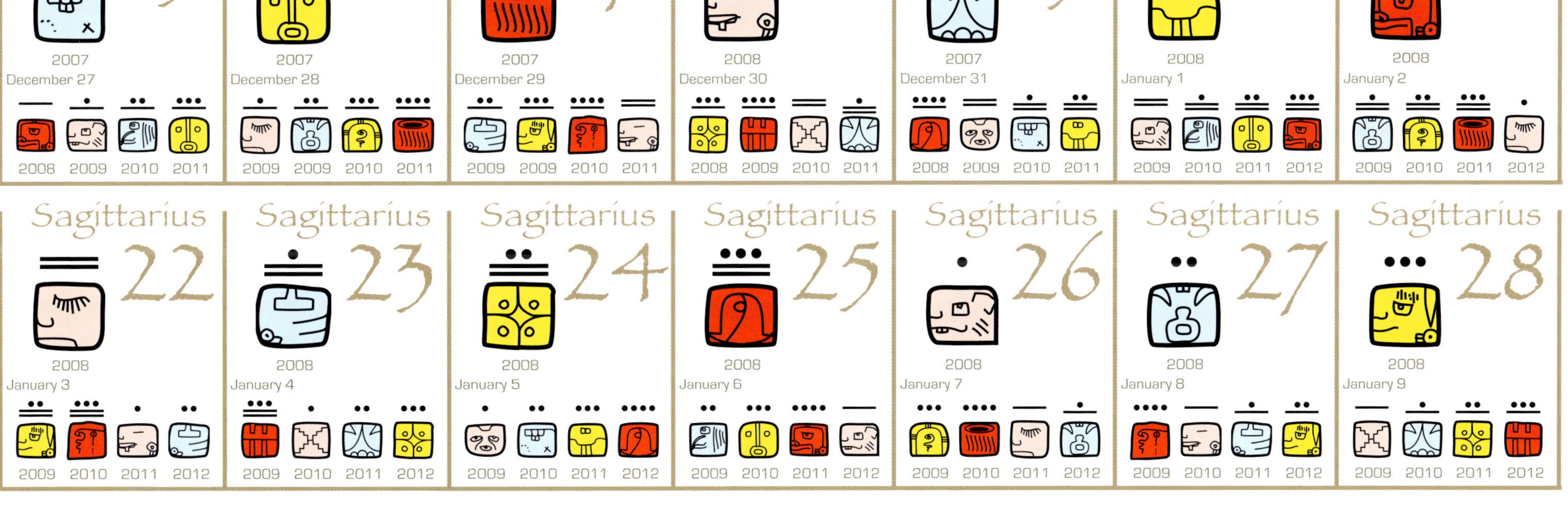

Sagittarius	Year	Date	Years
22	2008	January 3	2009 2010 2011 2012
23	2008	January 4	2009 2010 2011 2012
24	2008	January 5	2009 2010 2011 2012
25	2008	January 6	2009 2010 2011 2012
26	2008	January 7	2009 2010 2011 2012
27	2008	January 8	2009 2010 2011 2012
28	2008	January 9	2009 2010 2011 2012

MOON

Capricorn

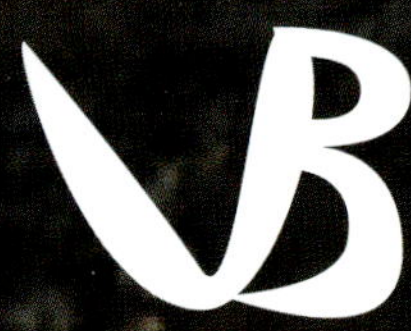

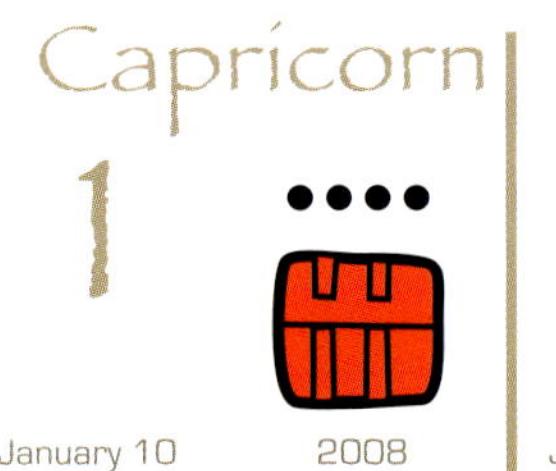
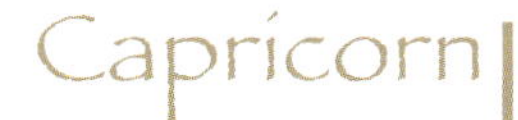
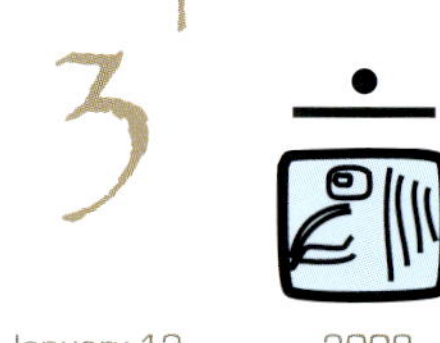
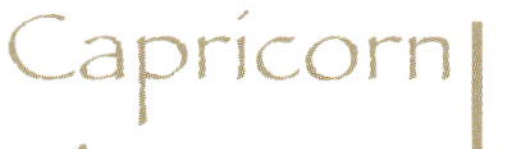
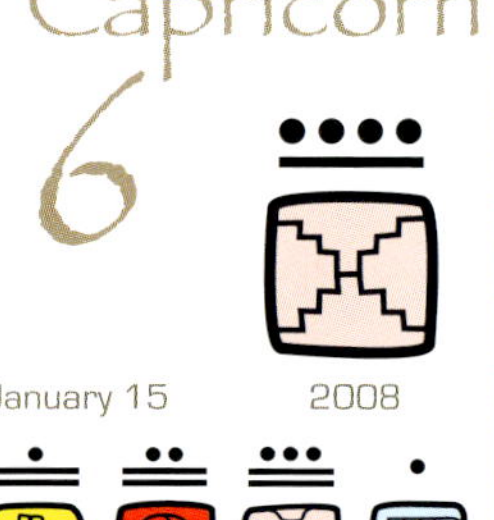
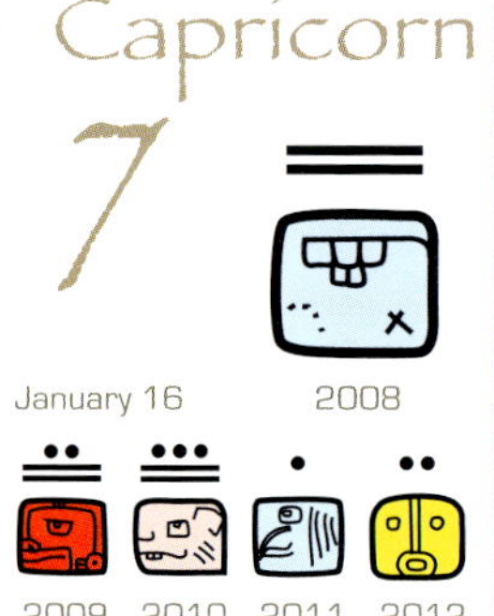

Capricorn	Capricorn	Capricorn	Capricorn	Capricorn	Capricorn	Capricorn
1	2	3	4	5	6	7
January 10 2008	January 11 2008	January 12 2008	January 13 2008	January 14 2008	January 15 2008	January 16 2008
2009 2010 2011 2012	2009 2010 2011 2012	2009 2010 2011 2012	2009 2010 2011 2012	2009 2010 2011 2012	2009 2010 2011 2012	2009 2010 2011 2012

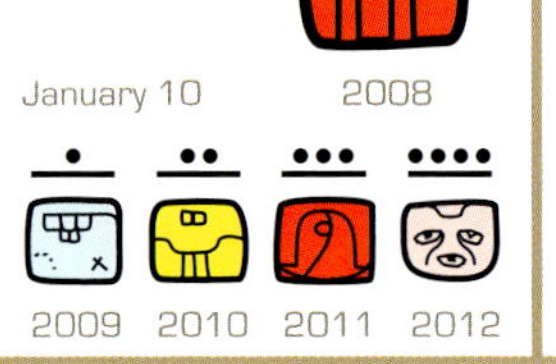
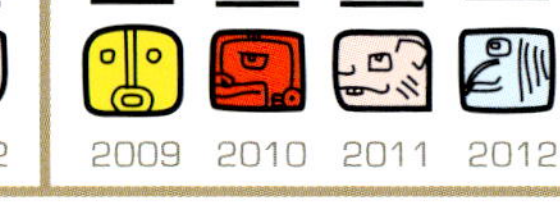
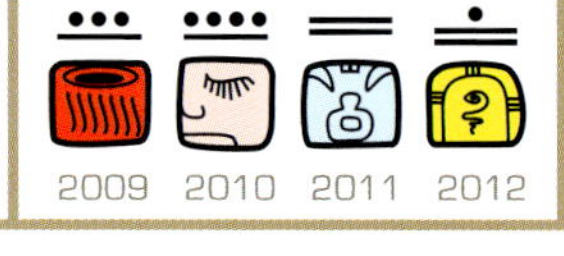
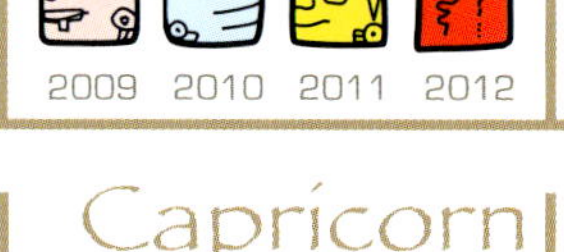
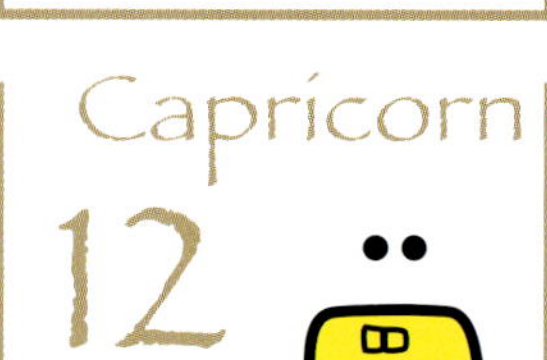

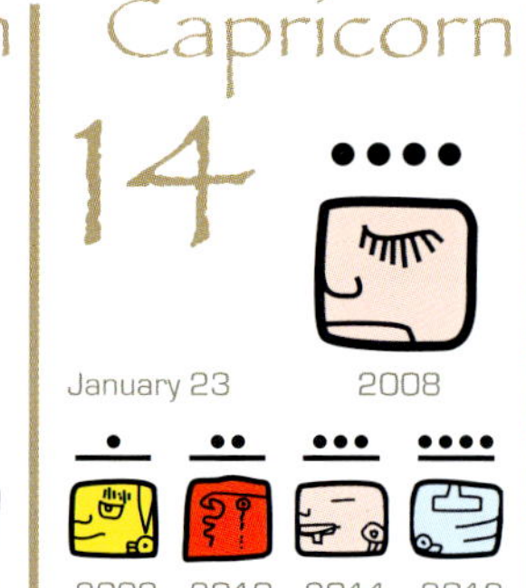
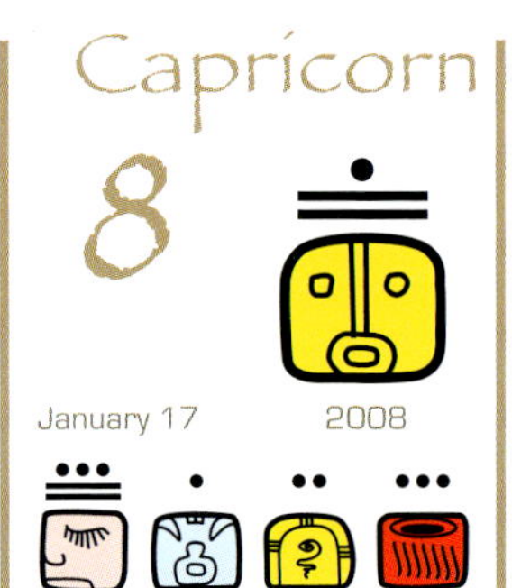

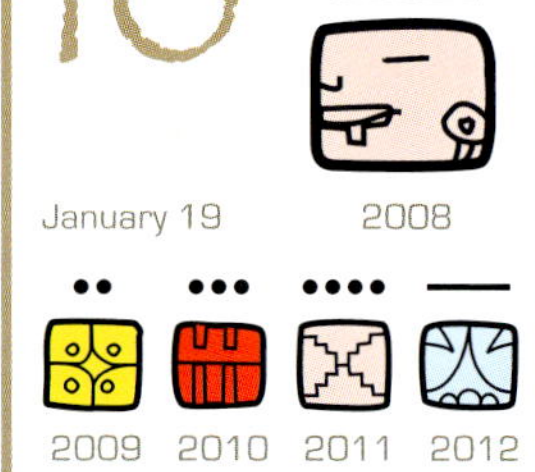
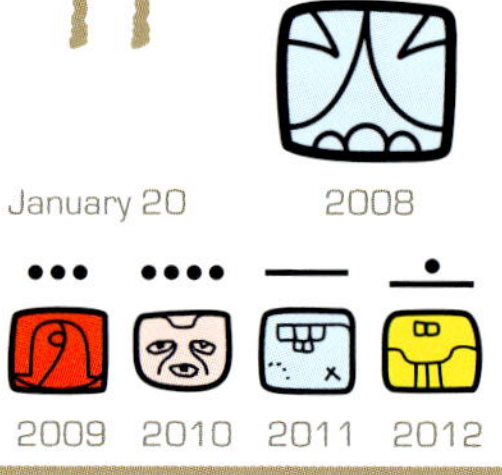
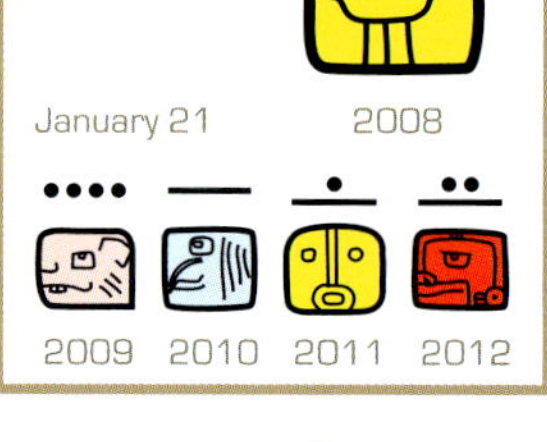
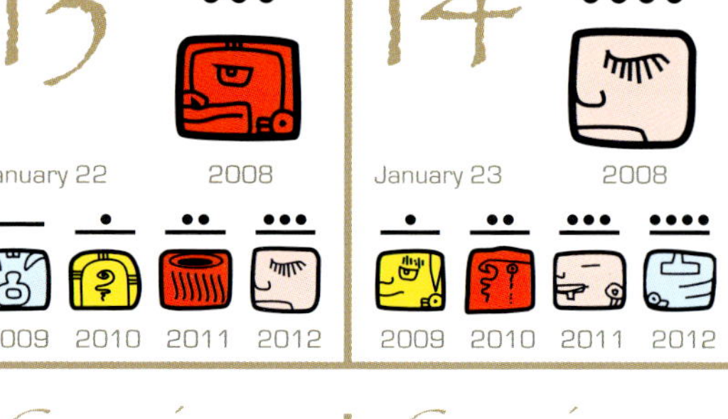

Capricorn	Capricorn	Capricorn	Capricorn	Capricorn	Capricorn	Capricorn
8	9	10	11	12	13	14
January 17 2008	January 18 2008	January 19 2008	January 20 2008	January 21 2008	January 22 2008	January 23 2008
2009 2010 2011 2012	2009 2010 2011 2012	2009 2010 2011 2012	2009 2010 2011 2012	2009 2010 2011 2012	2009 2010 2011 2012	2009 2010 2011 2012

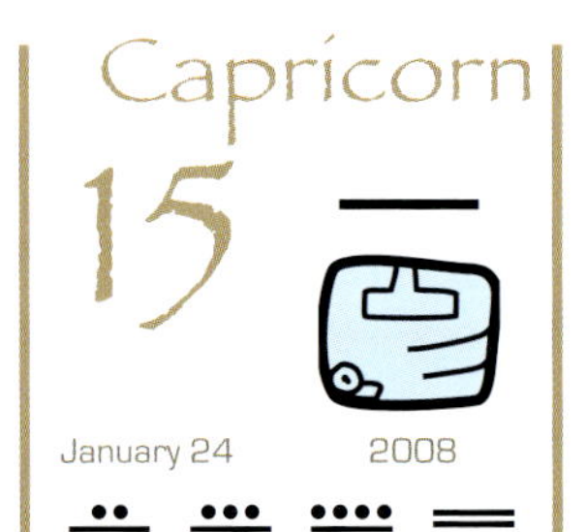

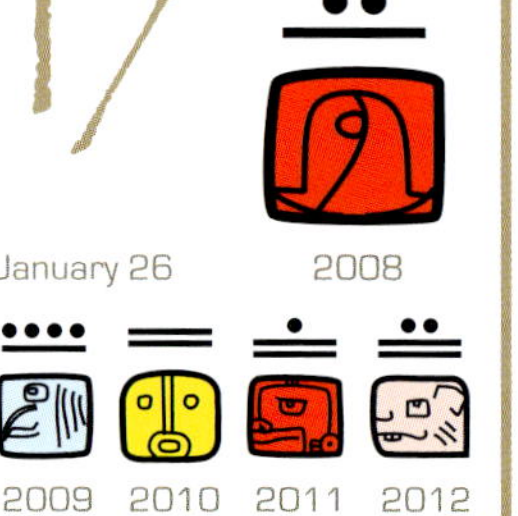
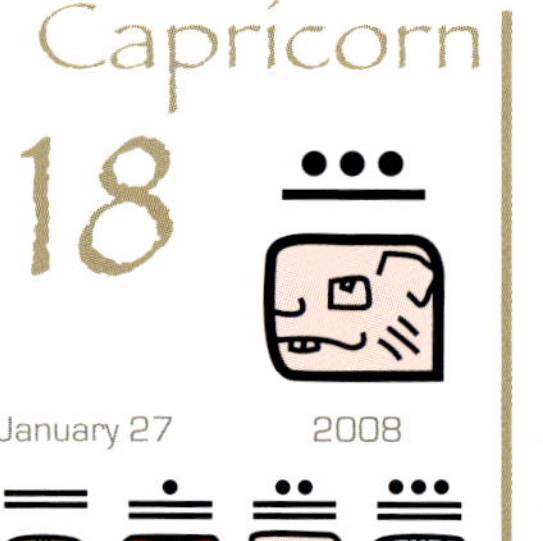
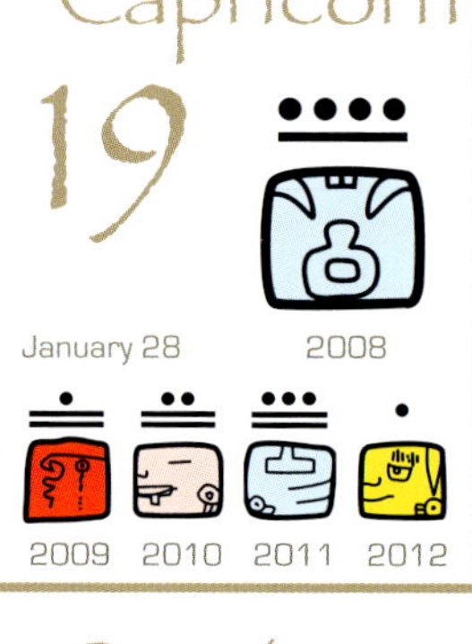
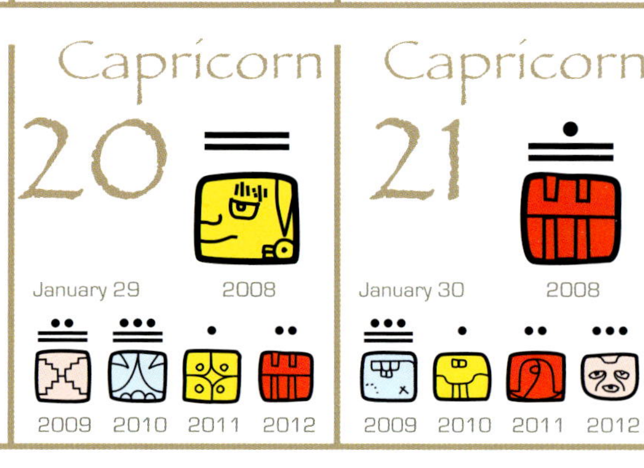

Capricorn	Capricorn	Capricorn	Capricorn	Capricorn	Capricorn	Capricorn
15	16	17	18	19	20	21
January 24 2008	January 25 2008	January 26 2008	January 27 2008	January 28 2008	January 29 2008	January 30 2008
2009 2010 2011 2012	2009 2010 2011 2012	2009 2010 2011 2012	2009 2010 2011 2012	2009 2010 2011 2012	2009 2010 2011 2012	2009 2010 2011 2012

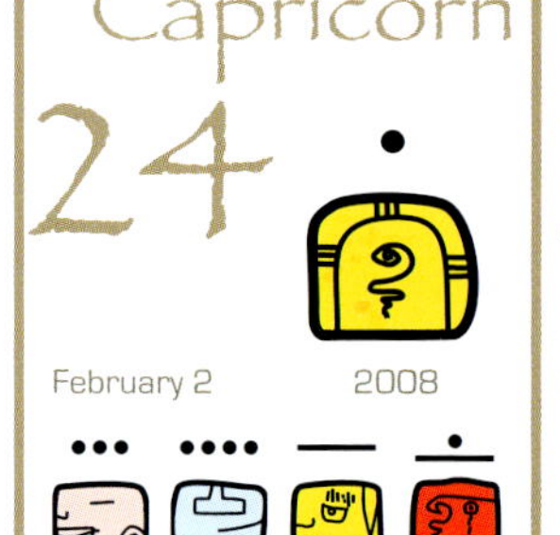
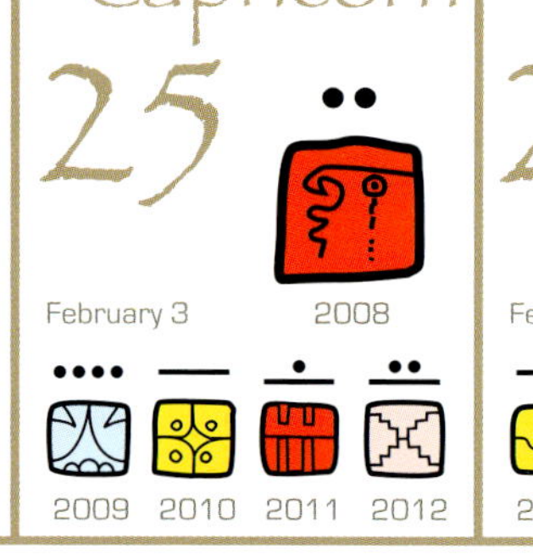
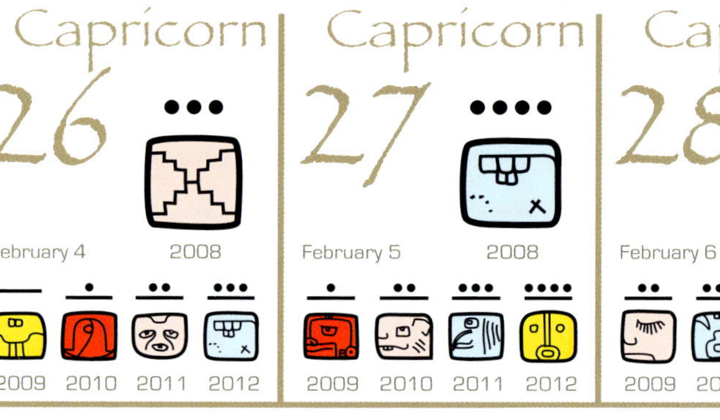
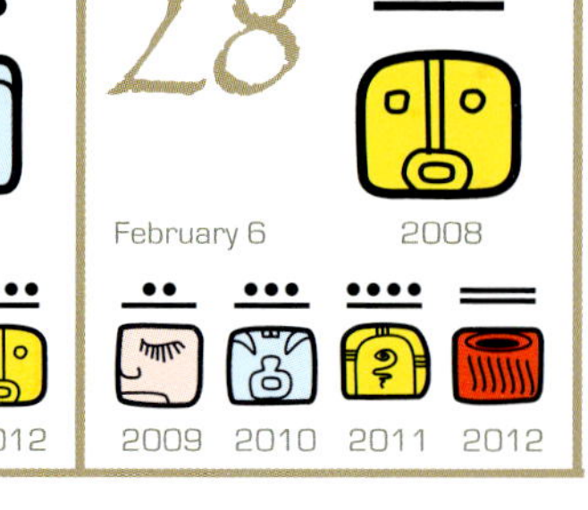

Capricorn	Capricorn	Capricorn	Capricorn	Capricorn	Capricorn	Capricorn
22	23	24	25	26	27	28
January 31 2008	February 1 2008	February 2 2008	February 3 2008	February 4 2008	February 5 2008	February 6 2008
2009 2010 2011 2012	2009 2010 2011 2012	2009 2010 2011 2012	2009 2010 2011 2012	2009 2010 2011 2012	2009 2010 2011 2012	2009 2010 2011 2012

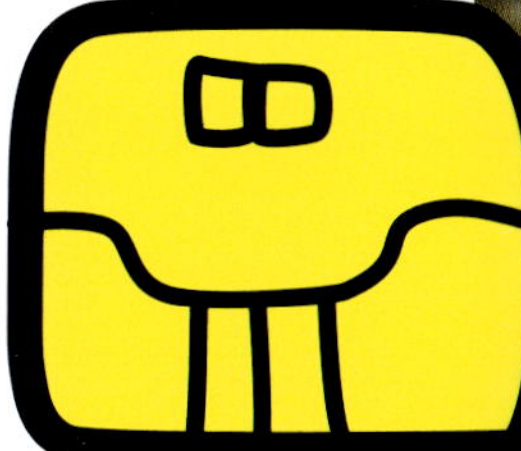

Kan means variously 'seed','corn' and 'lizard' - all are widespread representatives of fecundity, summer and the south. Kan is strongly associated with the physical realm and the power of renewal and growth. Kan is observation of and reverence for that which is seen to work in the natural world. planet ~ Jupiter. Affirmation: "I am the fertile ground and the self-germinating seed of possibility."

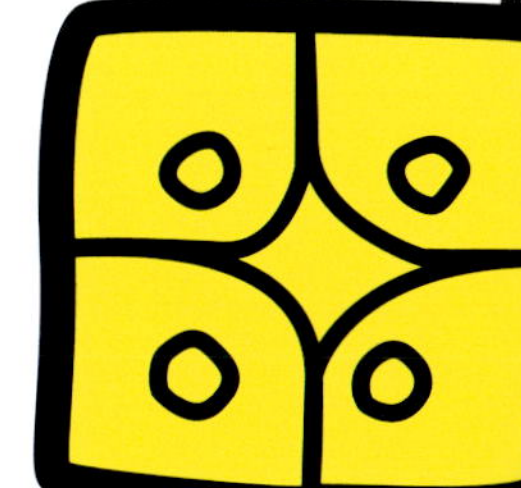

Lamat means 'star' or 'rabbit' and its primary association is rebirth and renewal. Lamat is a glyph of beauty and vitality and is the physical elegance: - the art - of transformation and recreation. The rabbit appreciates the fruits of the world created for him by the gods. planet ~ Venus. Affirmation: "I am my full presence now. I am the harmony of the stars."

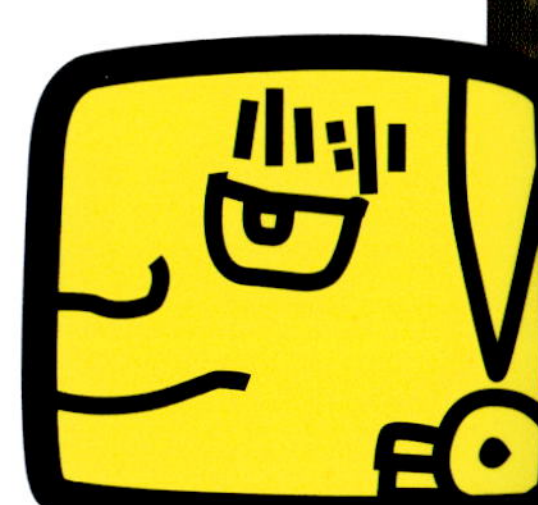

Eb means variously 'human'. Eb is associated with the sensual and physical aspects of living. Eb is the road of life and, in some manifestations, the greening and enlightening of individual and planetary awareness. Eb is free will it is consciousness walking in beauty and balance. All earthly kind belong to this road. Eb is individual spirit and authority and stands but one step from commUnity. planet ~ Earth. Affirmation: "I am an open chalice. I am a joyfull expression of the abundance of the universe."

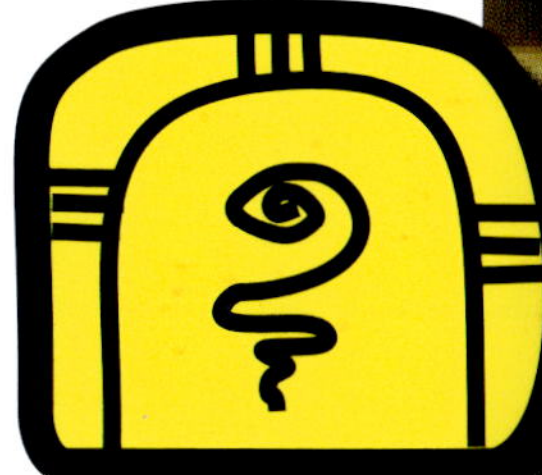

Cib means 'cosmic warrior' or 'vulture'. Cib is fearless and intelligent. Its primary resonance is of forgiveness and karmic catharsis. The transformic power of darkness embodies Cib. Cib is the successful movement from shadow to the light, the recognition and enactment of each ending as a new beginning. planet ~ Saturn. Affirmation: "I take up the staff of my power."

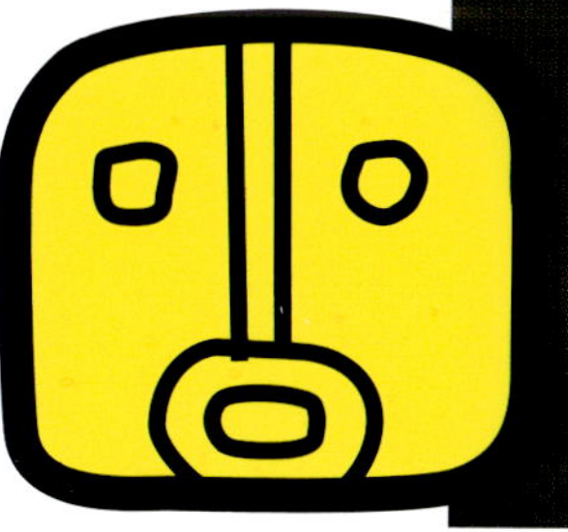

Ahau means 'Lord' or 'sun' and represents connection with our ancestral soul and with the epoch making events of both past and future. Ahau is completion Ahau is one of the twenty as well as an avatar of the twenty. The Quiche Maya envisage a 'sack of days' which is borne continually by each of us as we emerge from each new cycle of time. Ahau is a reminder of this continuity. planet ~ Pluto. Affirmation: "I am a child of the light, crowned with the self-authority and transparency of the divine child I know myself to be."

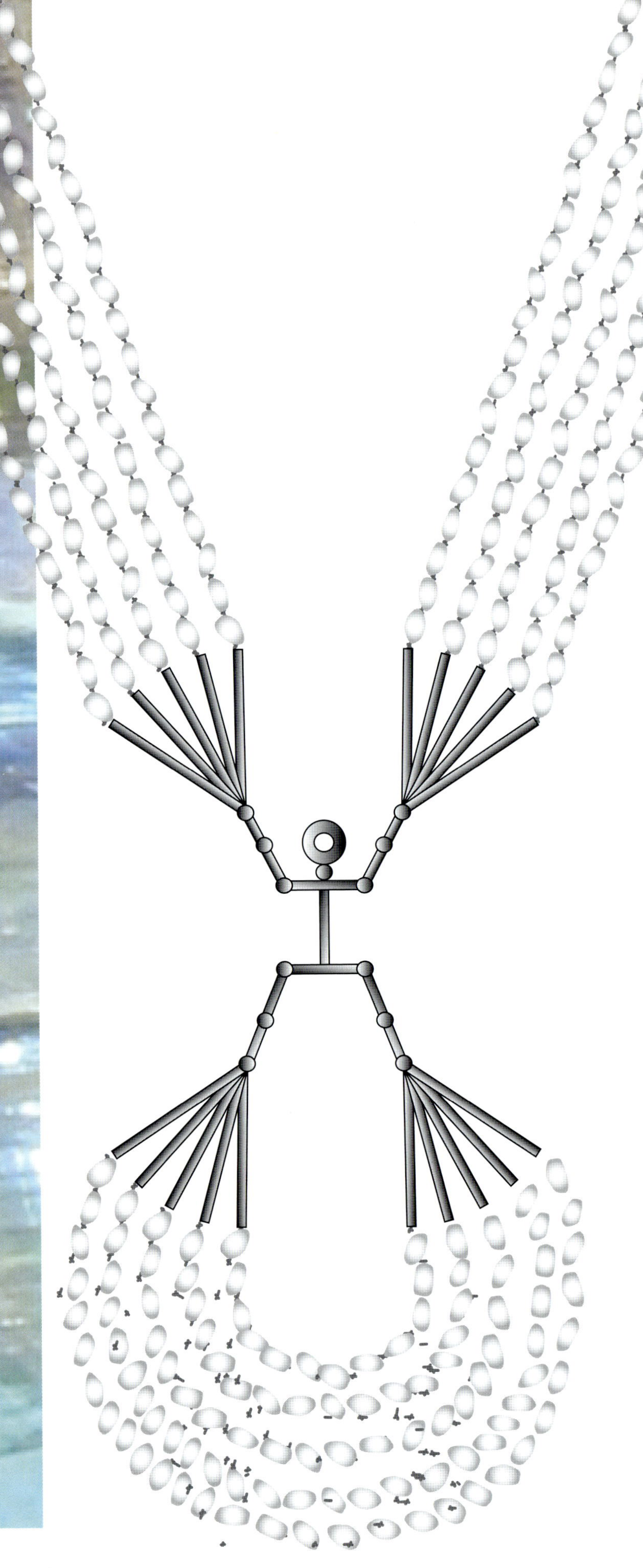

Numbers in Cosmic cycles and our body

The numbers in the Maya calendar show a resemblance between the cosmic cycle and the human body. It becomes 13 times full moon, we have 13 big joints which enable us to move around. Movement is time.

13 plays a crucial role in many ancient civilisations. It was considered holy in religious systems all over the world. In Christian religion, the number 13 played a noteworthy role in that Jesus and his disciples formed a group of thirteen members. The process of creation in the Bile takes place in 7 days. The Maya described the same process in 7 days and 6 nights which is 13.

Every holy number represents a stage in the creation process, it holds an intention and pulses creative energy in an upwards spiral. According to Isaac Newton energy is linear, Albert Einstein considered energy as curved. The traditional Maya described energy as waves. Each tone builds on the previous one, they are interwoven. The 13 tones of the creative impulse have value for different time frames, such as: 13 seconds, 13 days, 13 weeks, 13 months, 13 years.

MOON

Aquarius

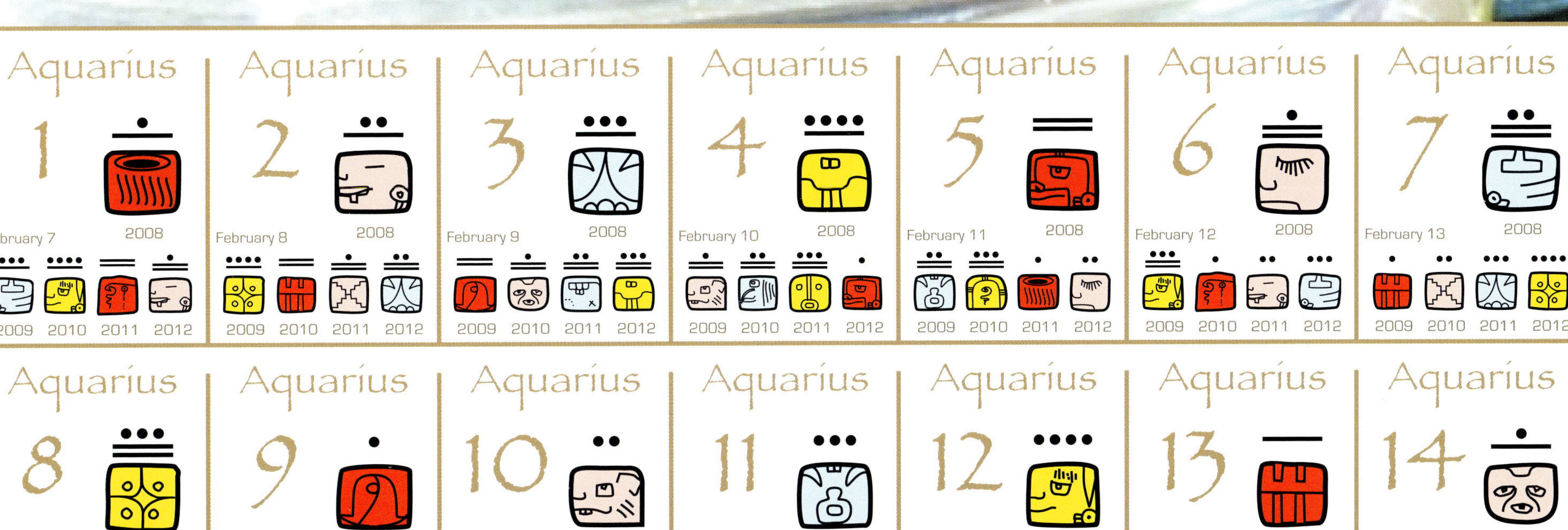

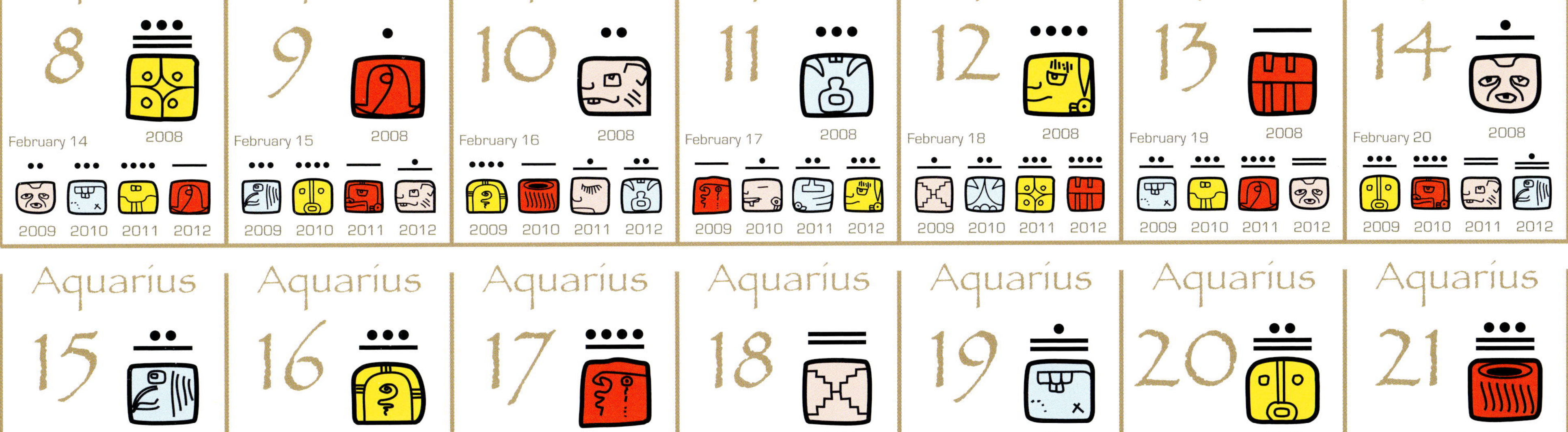

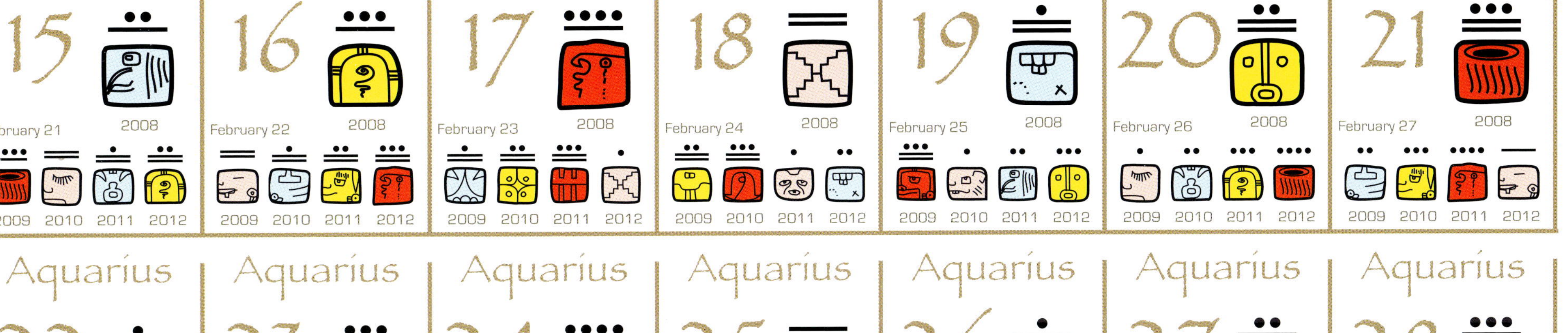

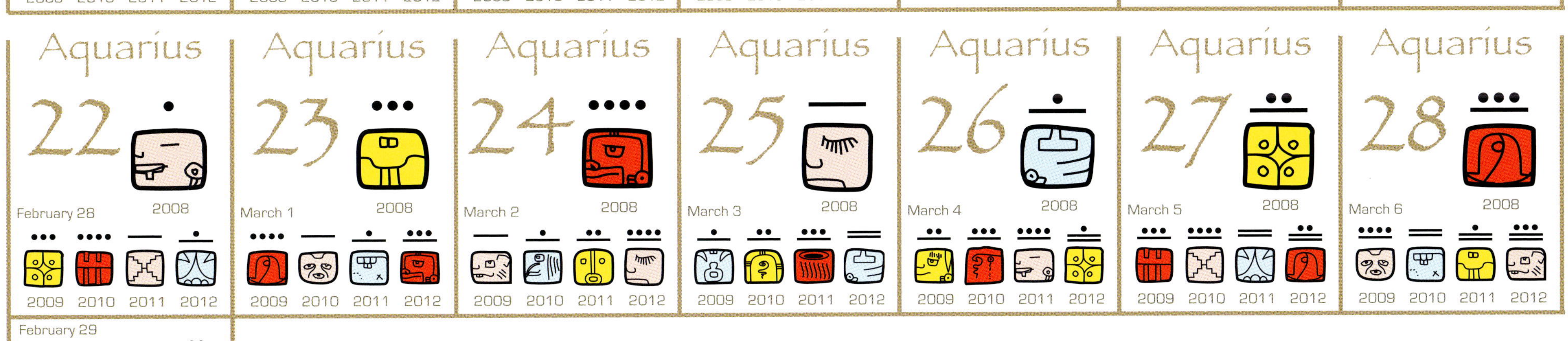

February 29
2008 2012

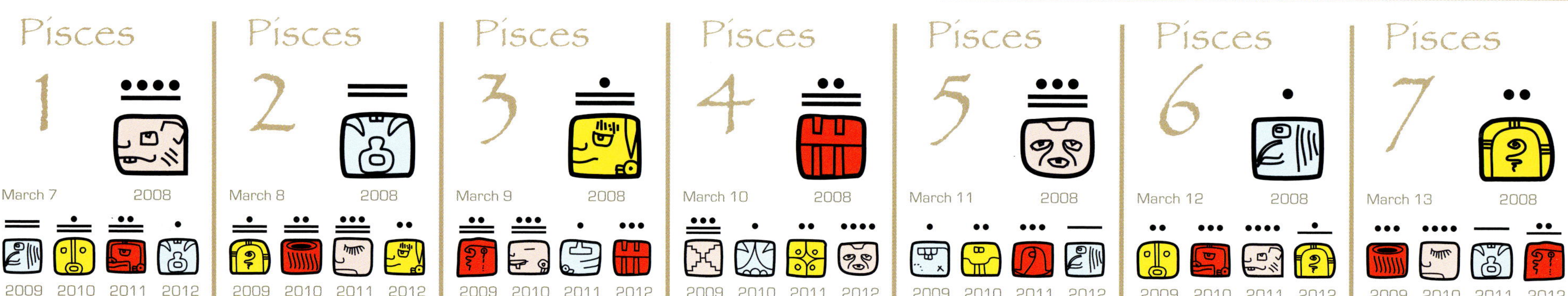
MOON
Pisces

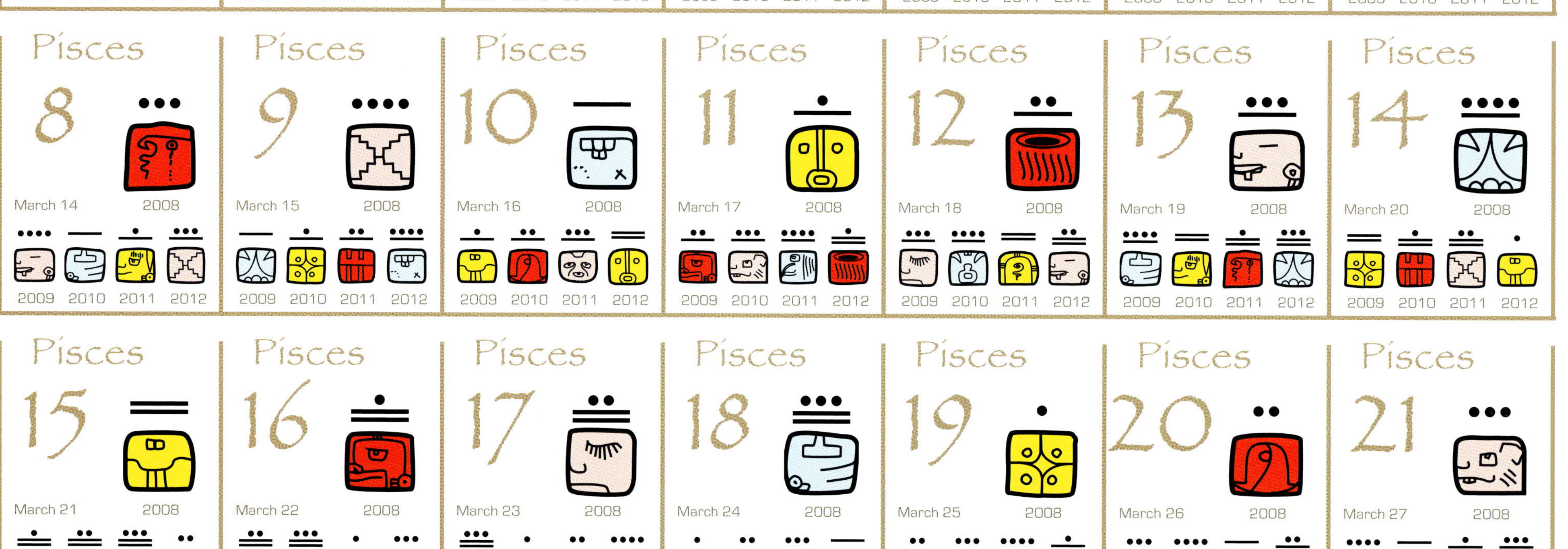
Pisces 1 March 7 2008 2009 2010 2011 2012
Pisces 2 March 8 2008 2009 2010 2011 2012
Pisces 3 March 9 2008 2009 2010 2011 2012
Pisces 4 March 10 2008 2009 2010 2011 2012
Pisces 5 March 11 2008 2009 2010 2011 2012
Pisces 6 March 12 2008 2009 2010 2011 2012
Pisces 7 March 13 2008 2009 2010 2011 2012
Pisces 8 March 14 2008 2009 2010 2011 2012
Pisces 9 March 15 2008 2009 2010 2011 2012
Pisces 10 March 16 2008 2009 2010 2011 2012
Pisces 11 March 17 2008 2009 2010 2011 2012
Pisces 12 March 18 2008 2009 2010 2011 2012
Pisces 13 March 19 2008 2009 2010 2011 2012
Pisces 14 March 20 2008 2009 2010 2011 2012

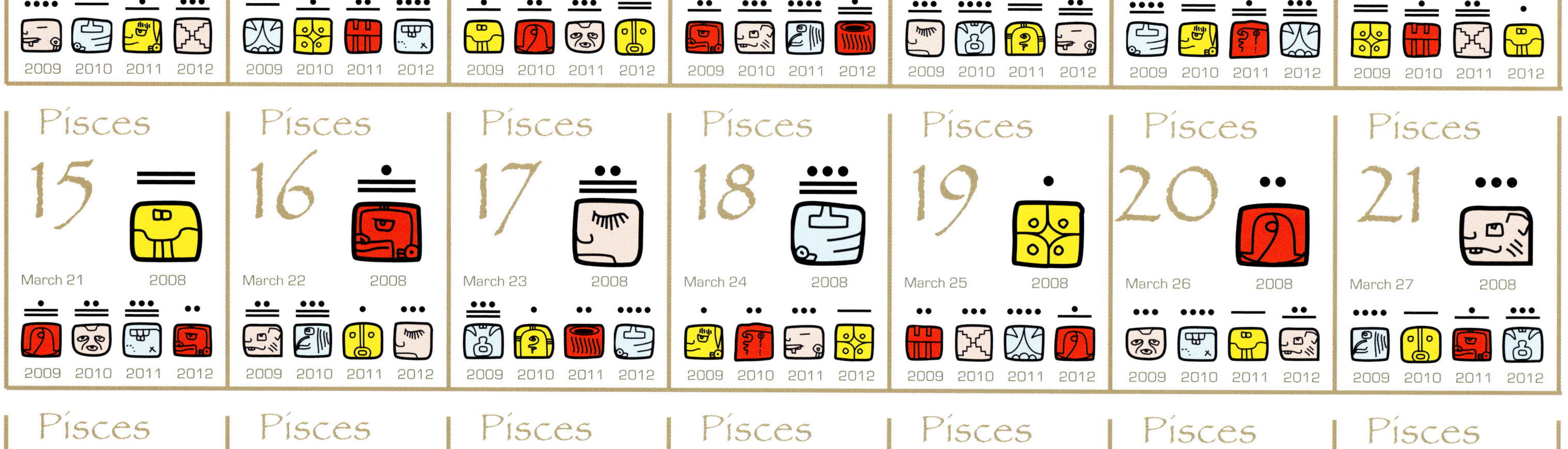
Pisces 15 March 21 2008 2009 2010 2011 2012
Pisces 16 March 22 2008 2009 2010 2011 2012
Pisces 17 March 23 2008 2009 2010 2011 2012
Pisces 18 March 24 2008 2009 2010 2011 2012
Pisces 19 March 25 2008 2009 2010 2011 2012
Pisces 20 March 26 2008 2009 2010 2011 2012
Pisces 21 March 27 2008 2009 2010 2011 2012

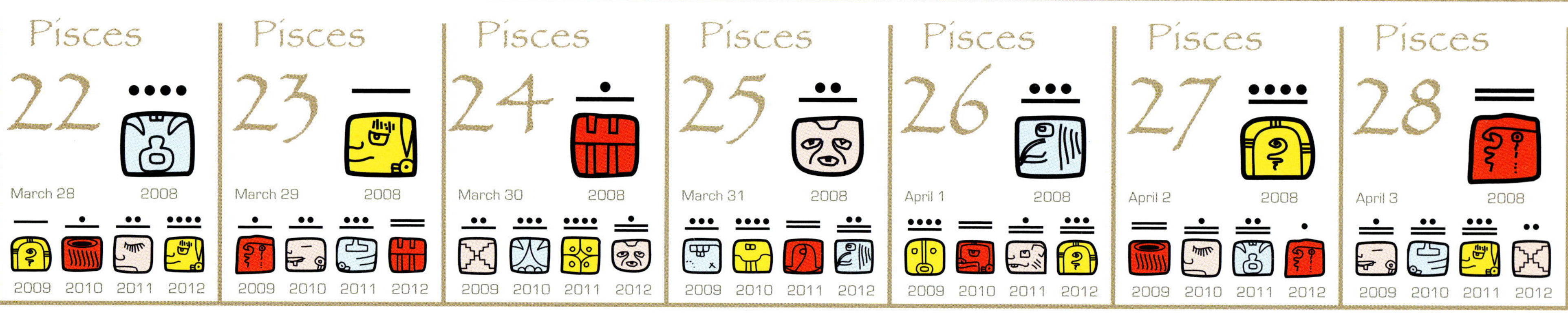
Pisces 22 March 28 2008 2009 2010 2011 2012
Pisces 23 March 29 2008 2009 2010 2011 2012
Pisces 24 March 30 2008 2009 2010 2011 2012
Pisces 25 March 31 2008 2009 2010 2011 2012
Pisces 26 April 1 2008 2009 2010 2011 2012
Pisces 27 April 2 2008 2009 2010 2011 2012
Pisces 28 April 3 2008 2009 2010 2011 2012

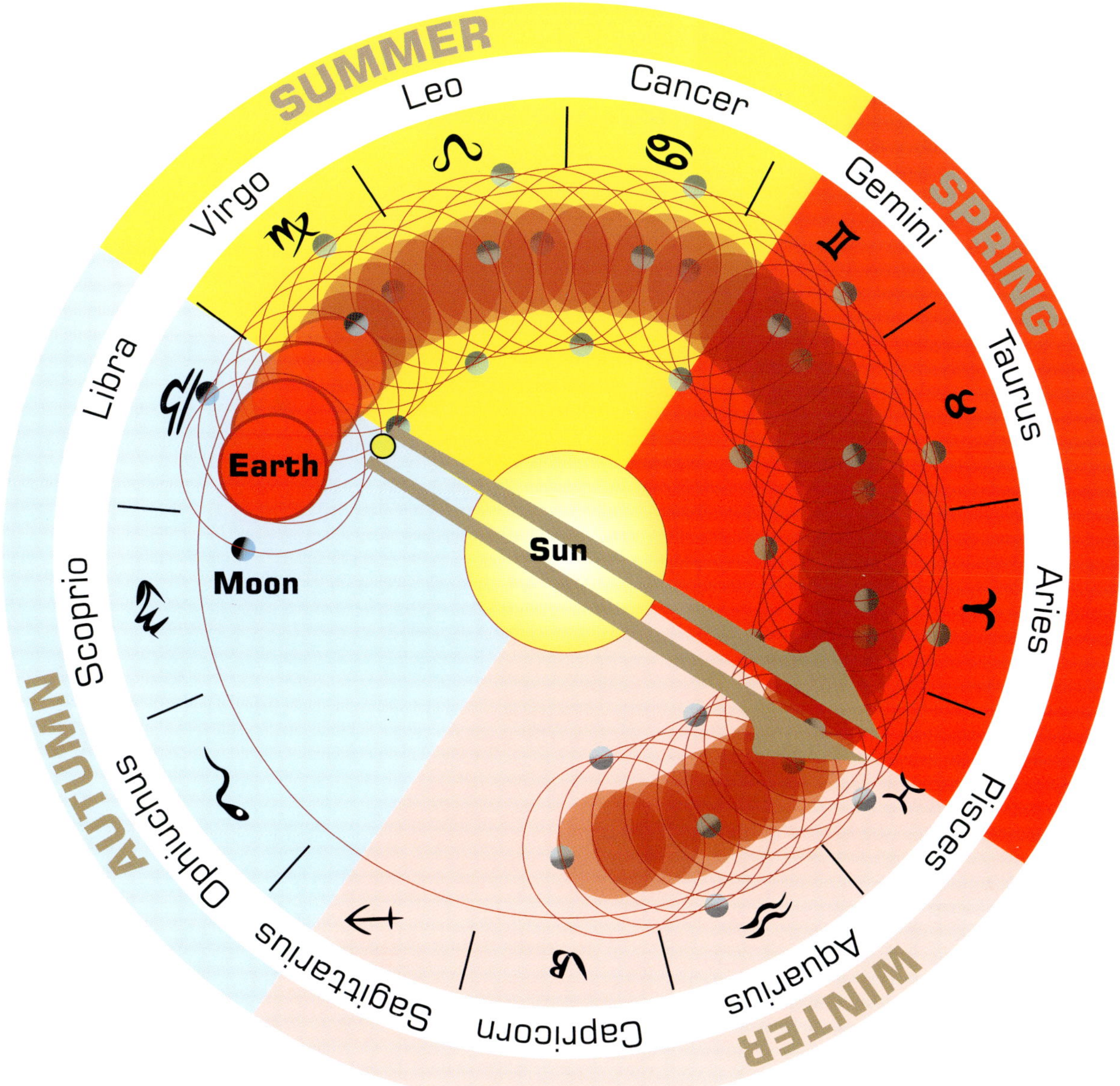

13 Months a Year

Annually it becomes much closer to 13 than 12 times full moon. Dividing a year in thirteen months is therefore more logical and makes the Moon easier to follow. The influence of the moon on all life on Earth is an ancient science, conquering grounds nowadays again. The sidereal month is the orbit of the Moon around Earth in 27,3 days. The Earth travels during a month a bit further around the Sun. The synodic month is time from full to full moon, which takes 29,5 days. 13 times 28 days is 364, which leaves one day spare. Every Monday starts at a 1st, a 8th, a 15th or a 22nd, so it will be easier to keep track of the days.

Naming the months

During the year when we travel on Earth counter clockwise around the Sun we "see" the Sun travelling through the signs of the zodiac. The 13 months are therefore named after the zodiac signs as commonly used in astrology. The familiarity of one's star~sign linked to the month of birth is far greater than the 23° difference in actual appearance. The first four weeks of the year are illustrated, above. We see our Sun in Leo, the hart of summer, yellow.

Ophiuchus

The 13th sign of the Zodiac is Ophiuchus, it holds the galactic centre, the Hunab Ku. Ophiuchus, the Snake Handler or Serpent Carrier brings new medicine to heal the world. Ophiuchus is traditionally identified with Aesculapius, who was so skilled in medicine and healing that it is said he could restore the dead. Ophiuchus is a big constellation, he stands with one foot on scorpio with Saggitarius on the other side.

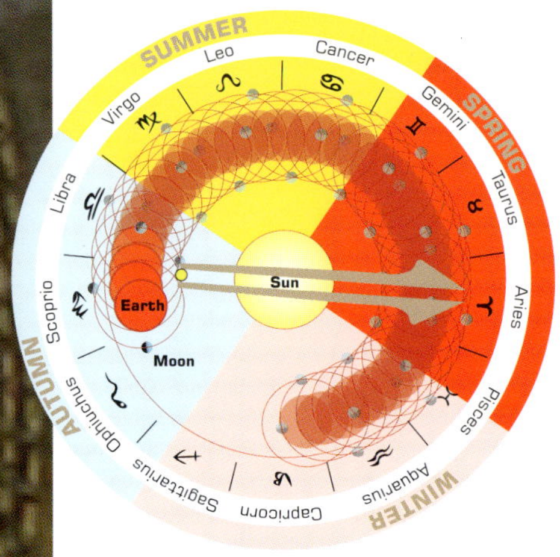

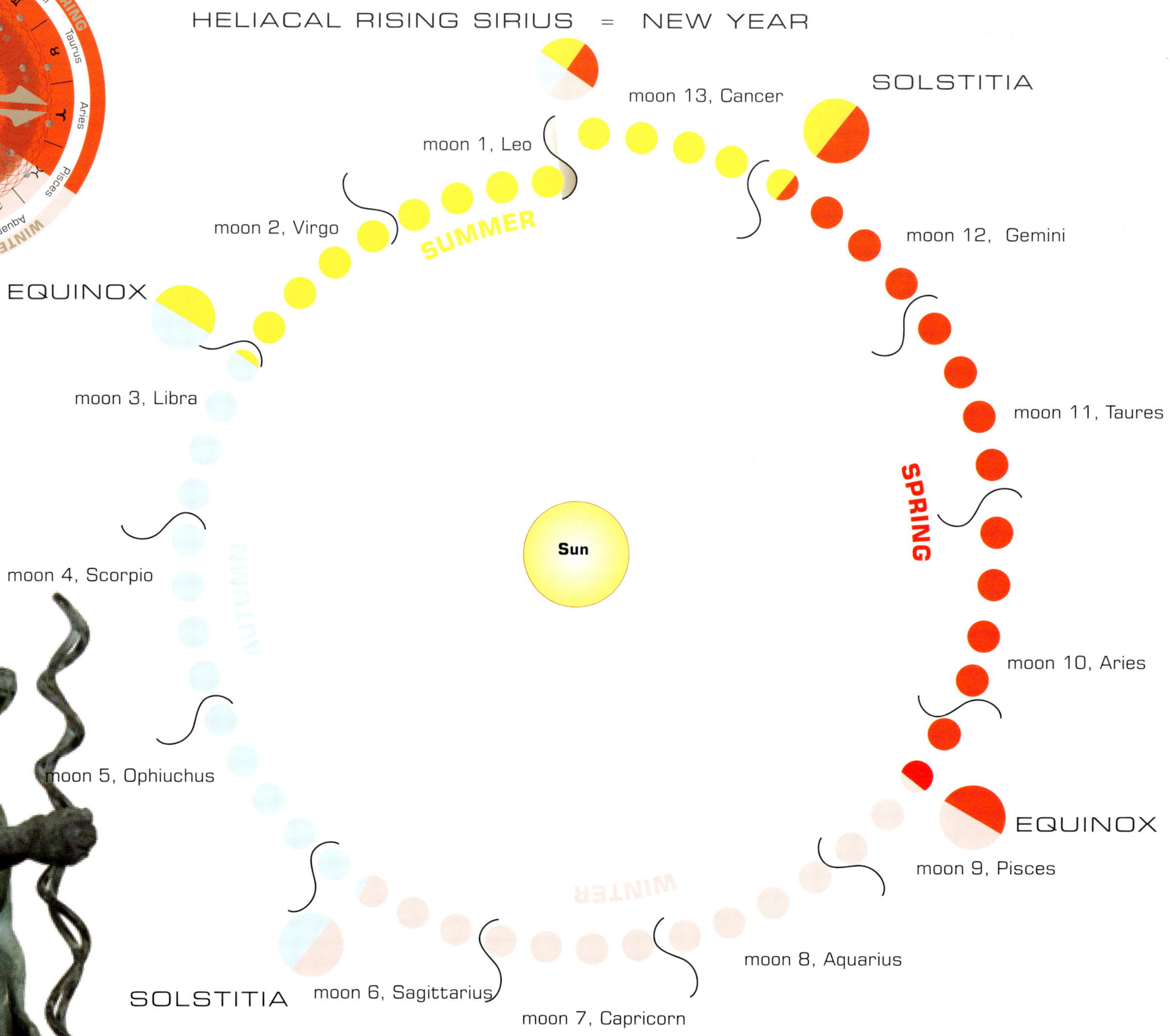

HELIACAL RISING SIRIUS = NEW YEAR

SOLSTITIA

moon 13, Cancer

moon 1, Leo

moon 2, Virgo

SUMMER

EQUINOX

moon 12, Gemini

moon 3, Libra

moon 11, Taures

SPRING

Sun

moon 4, Scorpio

AUTUMN

moon 10, Aries

moon 5, Ophiuchus

EQUINOX

moon 9, Pisces

WINTER

moon 8, Aquarius

SOLSTITIA

moon 6, Sagittarius

moon 7, Capricorn

New Year at Leo 1 = July 26

The heliacal rising of the star Sirius is the reason to start a new year at 26 July. Only once a year, Sirius and our Sun rise together in the morning twilight, at the eastern horizon. Sirius, the brightest star of the firmament, reappears after long absence. The difference in appearance is about 3 minutes, first Sirius than our Sun rises. 26 July is the only fixed point in the sky, yearly, which can be measured all over the globe. (Equinox shifts 1° every 72 years.) The heliacal rising of Sirius was also in use by the Egyptians. Our new year is also based on it; precisely half a year later when our Sun and Sirius are in opposition with each other.

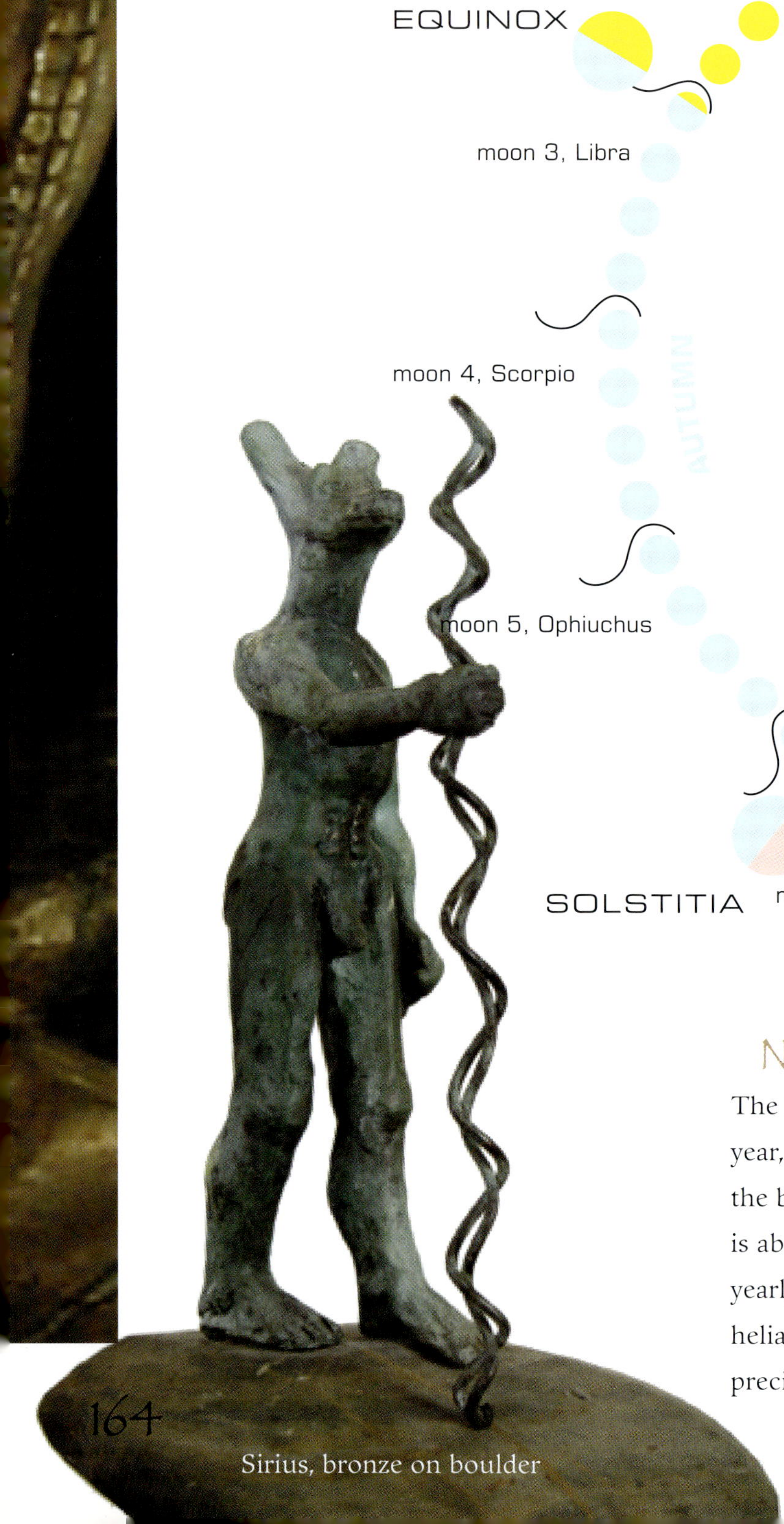

Sirius, bronze on boulder

MOON == Aries

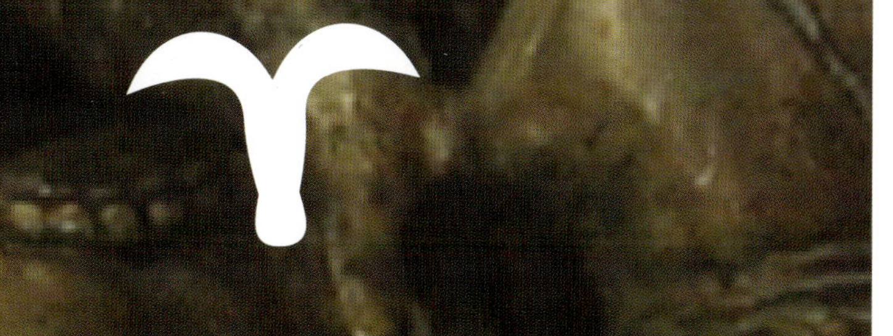

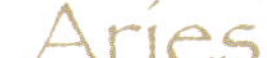 Aries 1 April 4 2008 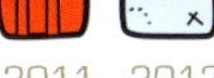2009 2010 2011 2012	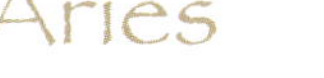 Aries 2 April 5 2008 2009 2010 2011 2012	Aries 3 April 6 2008 2009 2010 2011 2012	Aries 4 April 7 2008 2009 2010 2011 2012	Aries 5 April 8 2008 2009 2010 2011 2012	Aries 6 April 9 2008 2009 2010 2011 2012	Aries 7 April 10 2008 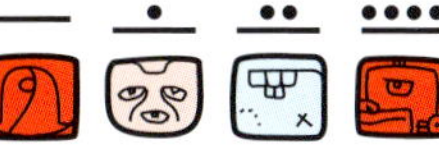2009 2010 2011 2012
Aries 8 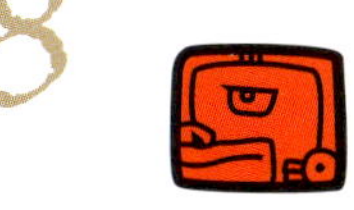April 11 2008 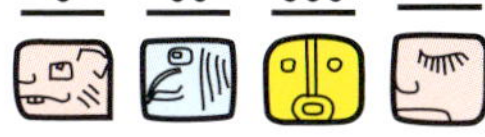2009 2010 2011 2012	Aries 9 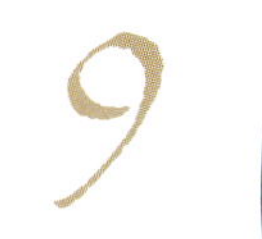April 12 2008 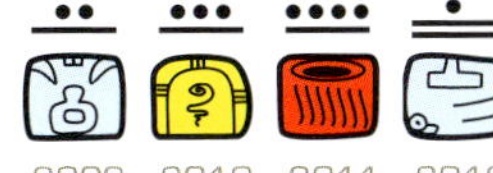2009 2010 2011 2012	Aries 10 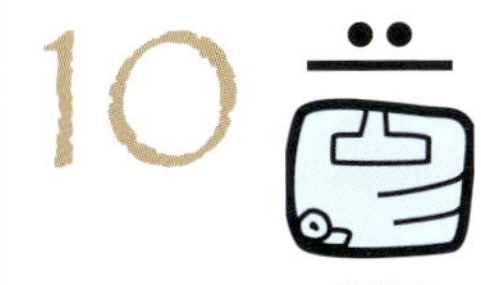April 13 2008 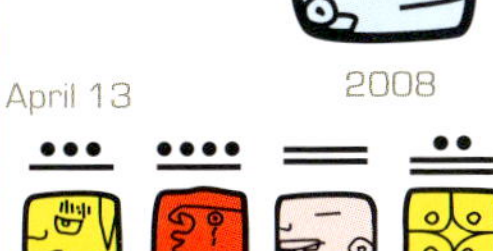2009 2010 2011 2012	Aries 11 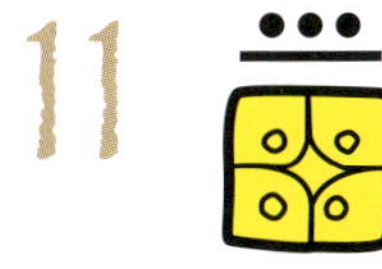April 14 2008 2009 2010 2011 2012	Aries 12 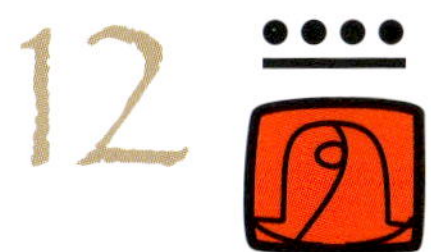April 15 2008 2009 2010 2011 2012	Aries 13 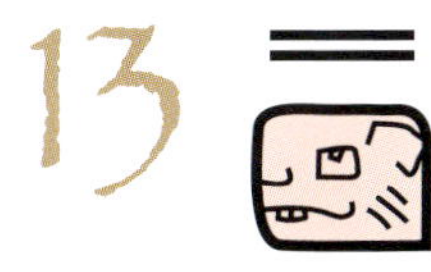April 16 2008 2009 2010 2011 2012	Aries 14 April 17 2008 2009 2010 2011 2012
Aries 15 April 18 2008 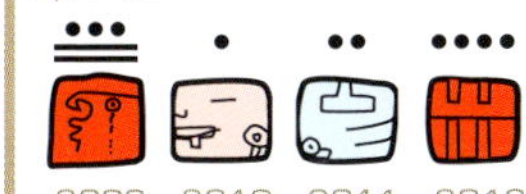2009 2010 2011 2012	Aries 16 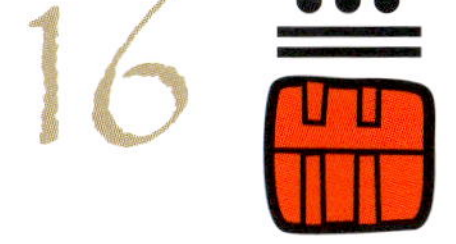April 19 2008 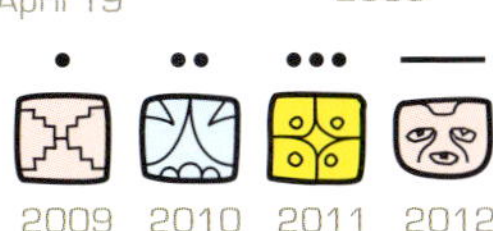2009 2010 2011 2012	Aries 17 April 20 2008 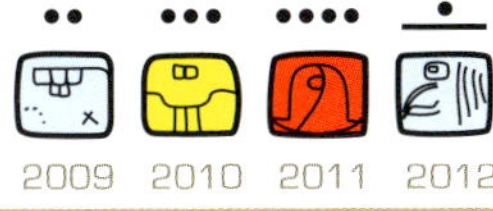2009 2010 2011 2012	Aries 18 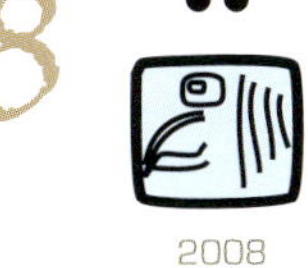April 21 2008 2009 2010 2011 2012	Aries 19 April 22 2008 2009 2010 2011 2012	Aries 20 April 23 2008 2009 2010 2011 2012	Aries 21 April 24 2008 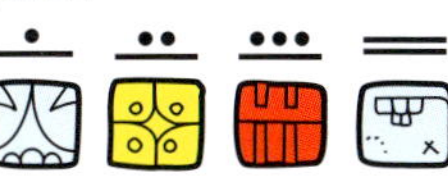2009 2010 2011 2012
Aries 22 April 25 2008 2009 2010 2011 2012	Aries 23 April 26 2008 2009 2010 2011 2012	Aries 24 April 27 2008 2009 2010 2011 2012	Aries 25 April 28 2008 2009 2010 2011 2012	Aries 26 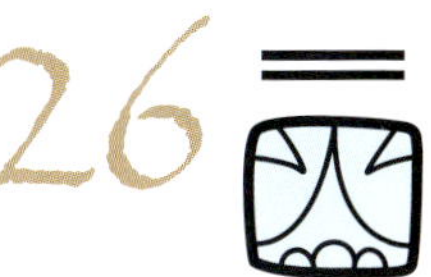April 29 2008 2009 2010 2011 2012	Aries 27 April 30 2008 2009 2010 2011 2012	Aries 28 May 1 2008 2009 2010 2011 2012

MOON

Taurus

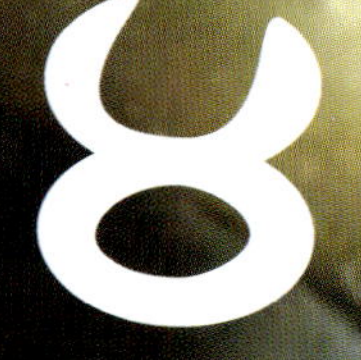

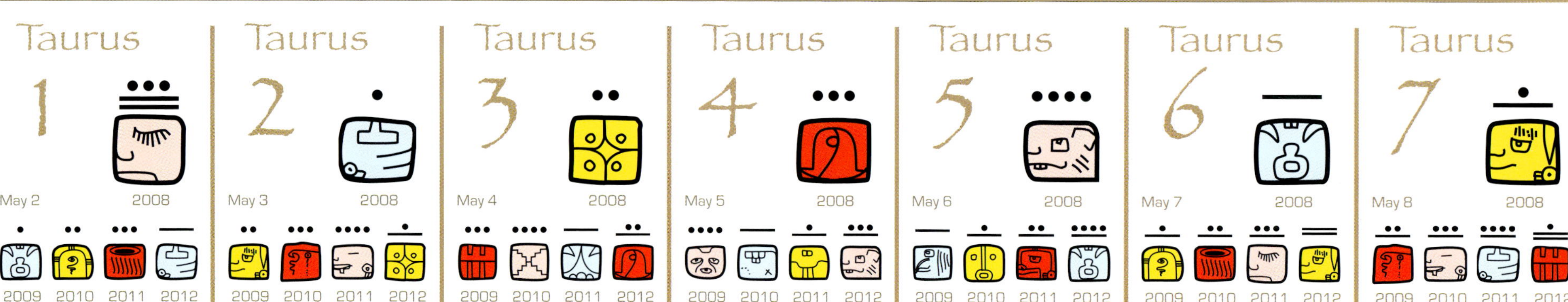

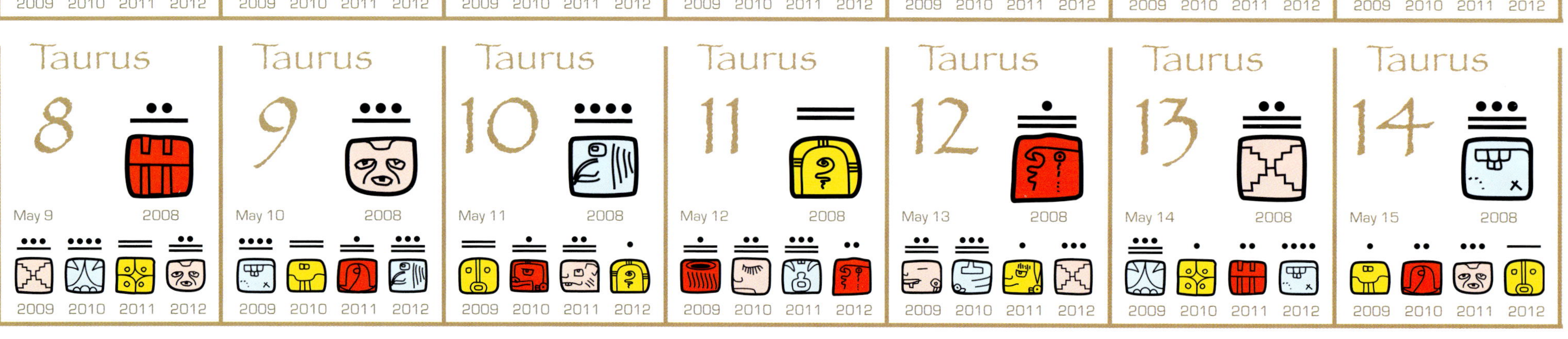

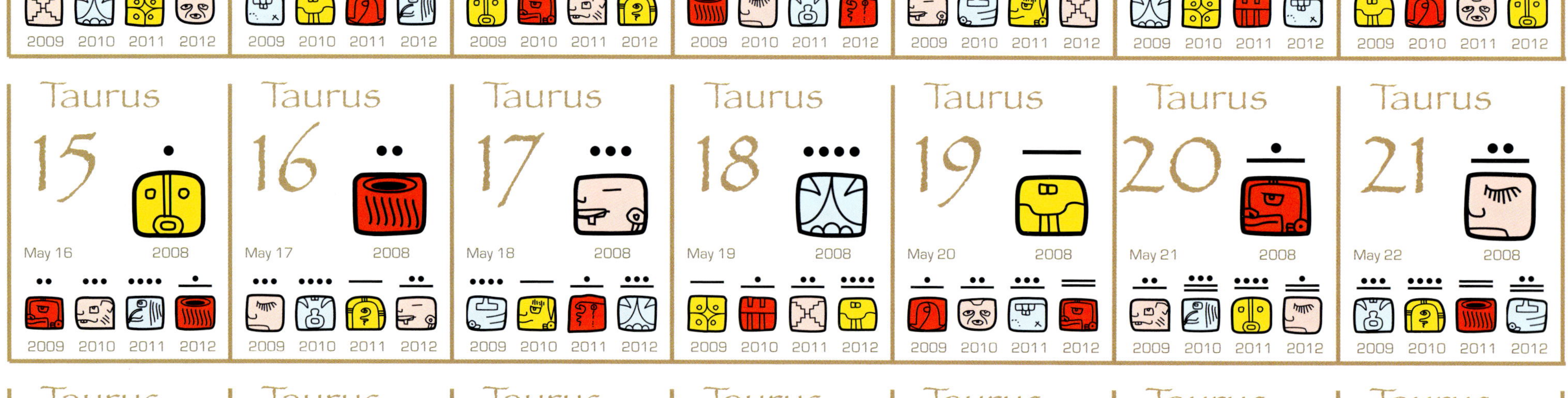

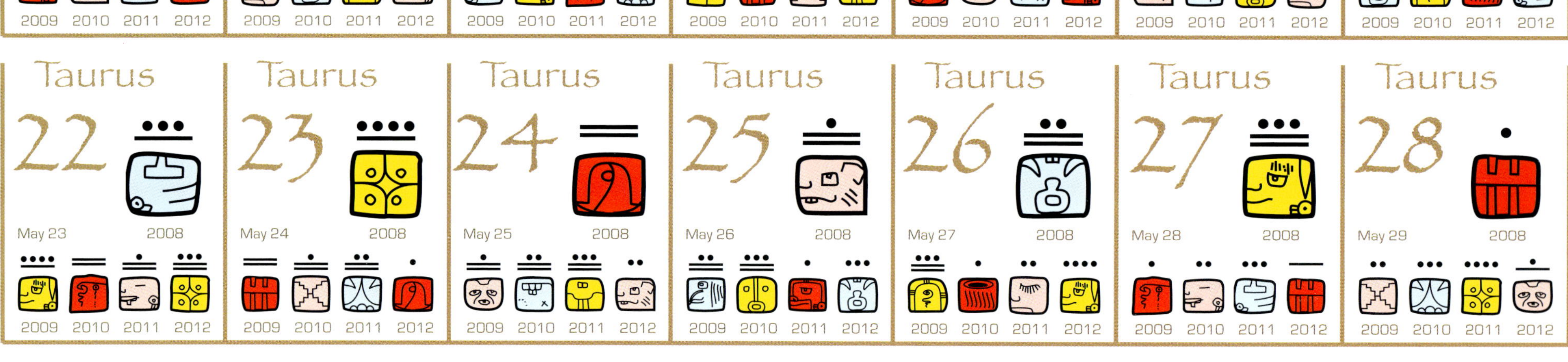

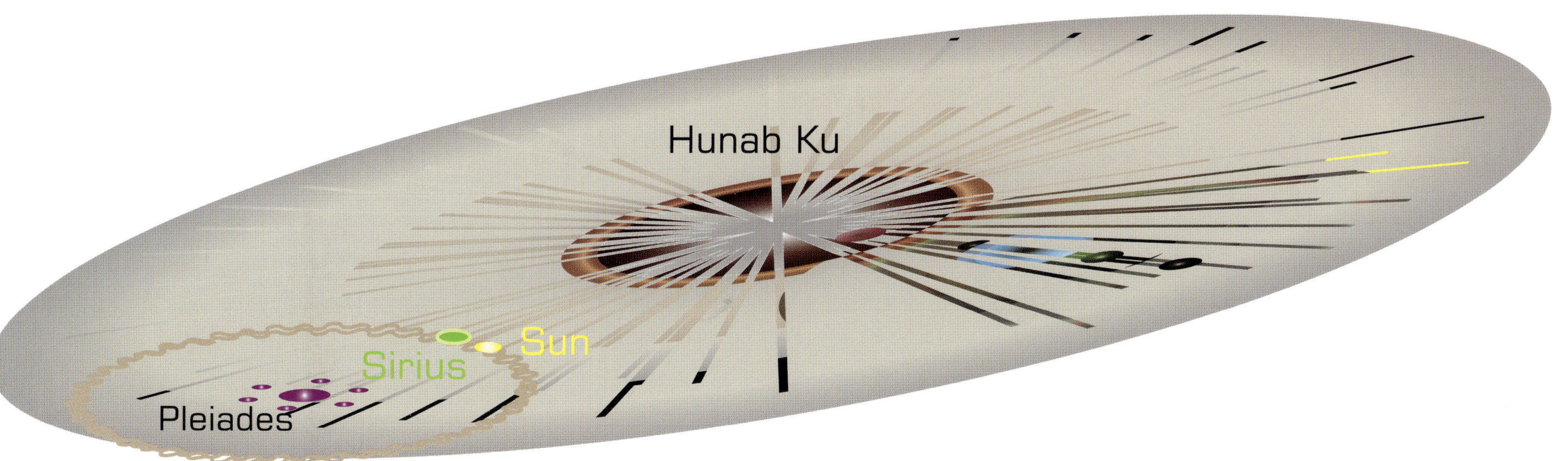

2012

The early Maya intended the 2012 end-date to target the rare alignment of the December solstice sun with the bright band of the Milky Way near the Galactic Center. The Galactic alignment is a rare astronomical event happening every 12,960 years; it does not signal the end of the world rather the beginning of an ascension stage in the development of human consciousness.

With the galactic alignment in 2012, many cycles will be completed. Sirius travels with the sun, like our DNA, orbiting the Pleiades in 26,000 years. Our solar system, the Pleiades and Sirius make a rotation around the Hunab Ku, in a period of 225 million years. Reptiles are the only species on Earth to fulfill this big cycle; they will only just make it. Now that the planet is becoming hotter, crocodiles may be of great help for human survival, because they are the beholders of the biological intelligence and kundalini energy running along their spines. 225 million years ago all continents were one unity: Pangaia.

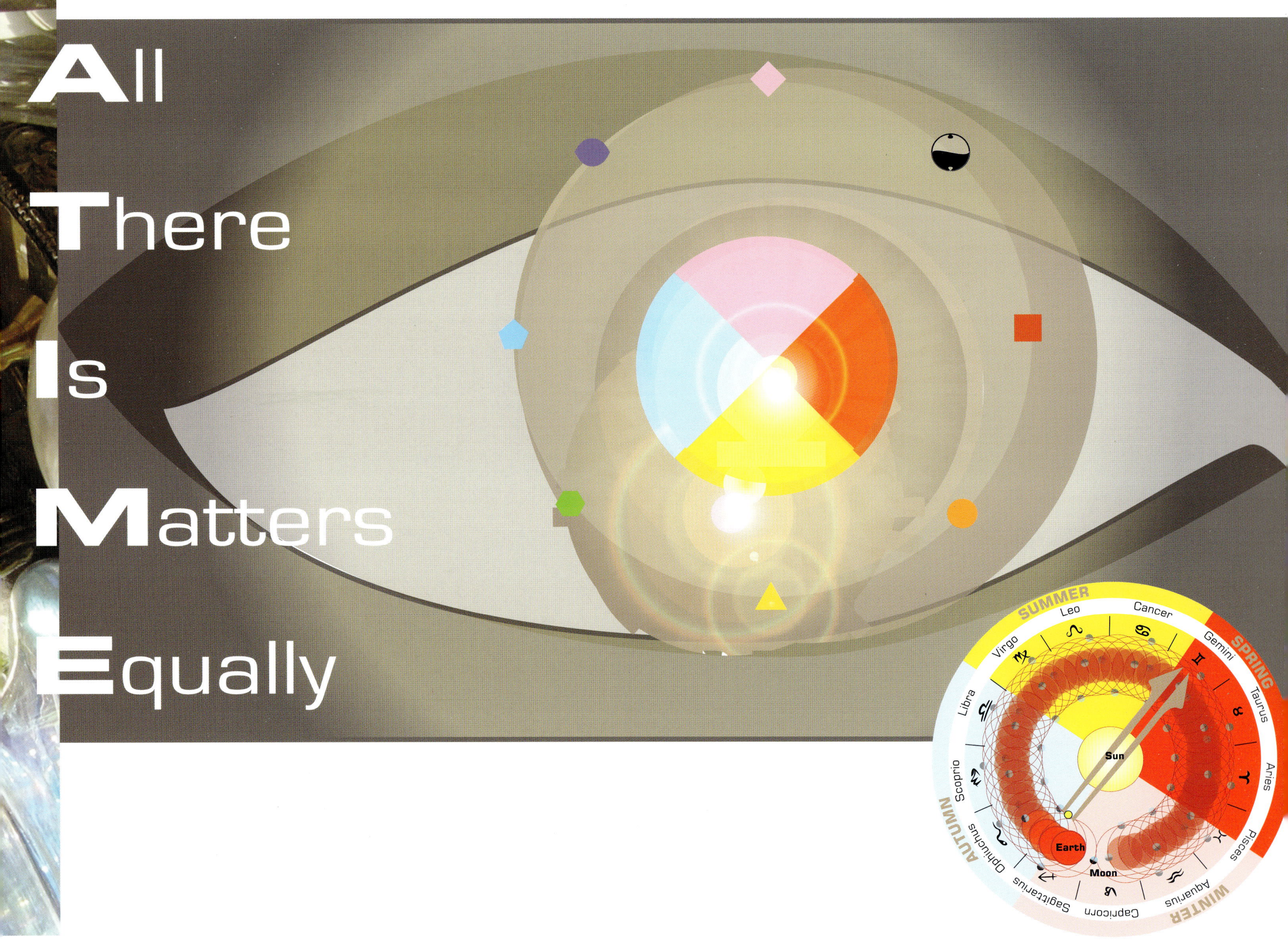
All
There
Is
Matters
Equally
SUMMER
Leo
Cancer
Gemini
SPRING
Taurus
Aries
Pisces
WINTER
Aquarius
Capricorn
Sagittarius
Ophiuchus
AUTUMN
Scoprio
Libra
Virgo
Sun
Earth
Moon

MOON Gemini

Gemini 1 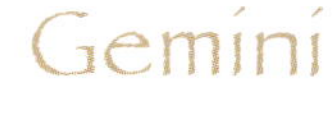 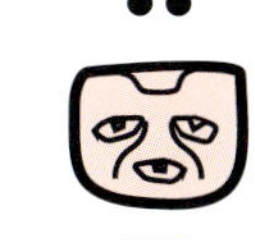 May 30 2008 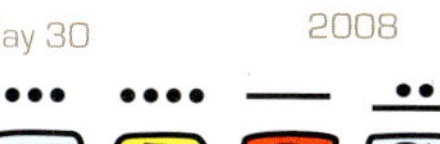 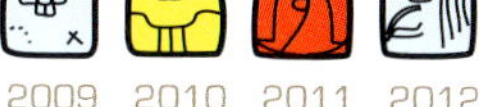 2009 2010 2011 2012	Gemini 2 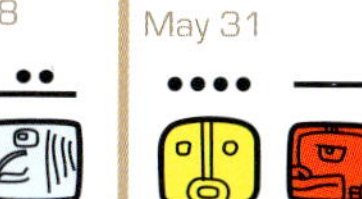 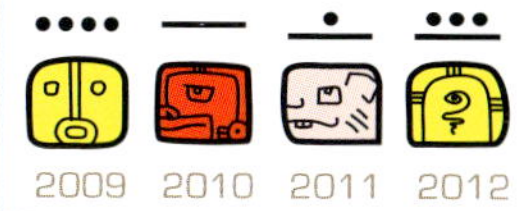 May 31 2008 2009 2010 2011 2012	Gemini 3 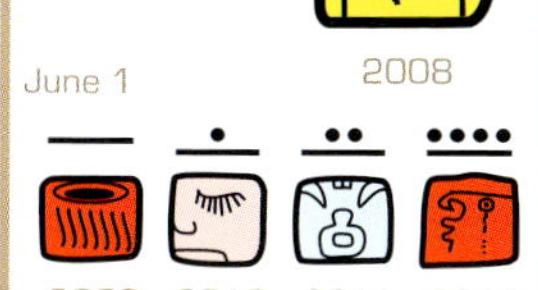 June 1 2008 2009 2010 2011 2012	Gemini 4 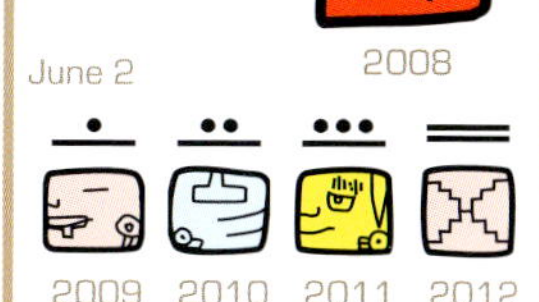 June 2 2008 2009 2010 2011 2012	Gemini 5 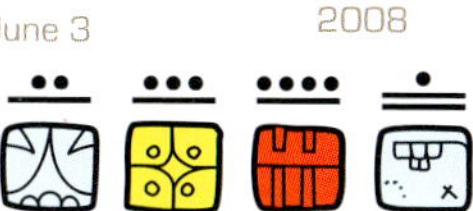 June 3 2008 2009 2010 2011 2012	Gemini 6 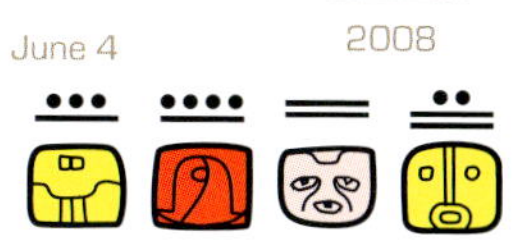 June 4 2008 2009 2010 2011 2012	Gemini 7 June 5 2008 2009 2010 2011 2012
Gemini 8 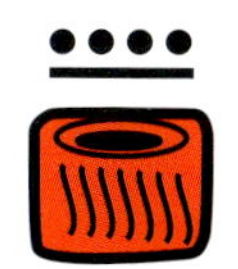 June 6 2008 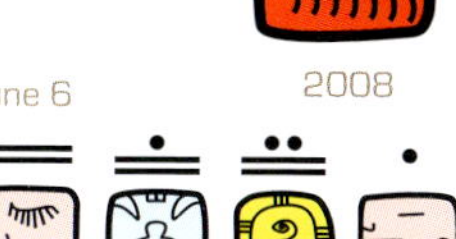 2009 2010 2011 2012	Gemini 9 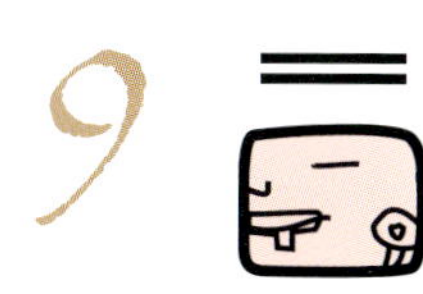 June 7 2008 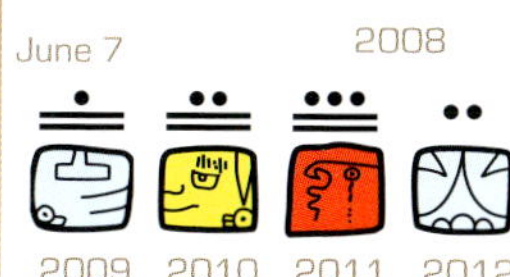 2009 2010 2011 2012	Gemini 10 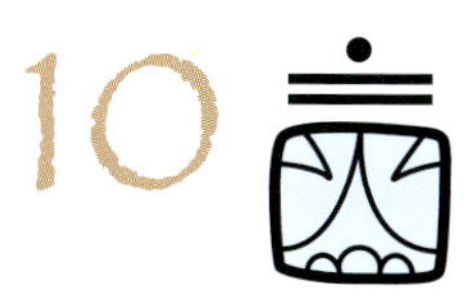 June 8 2008 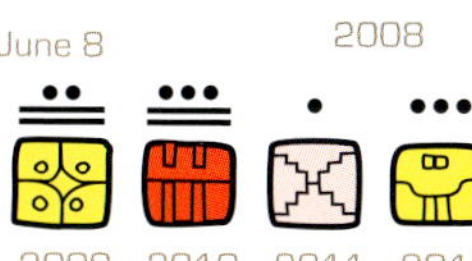 2009 2010 2011 2012	Gemini 11 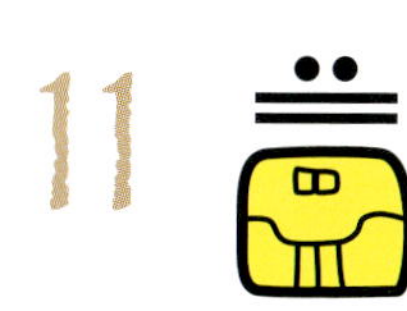 June 9 2008 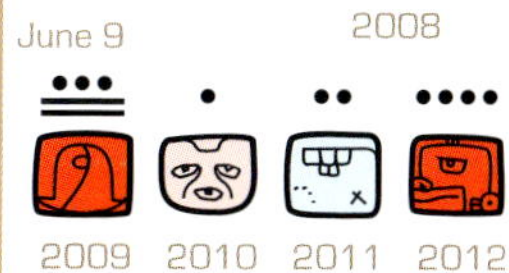 2009 2010 2011 2012	Gemini 12 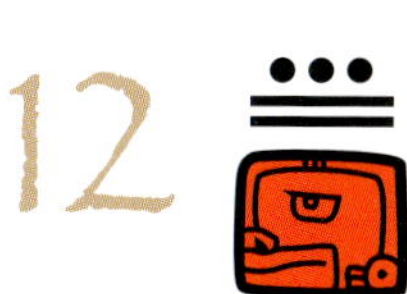 June 10 2008  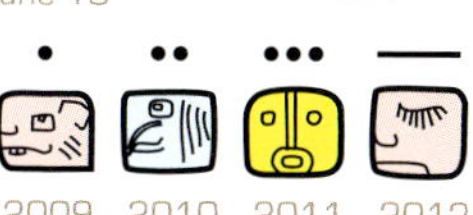 2009 2010 2011 2012	Gemini 13 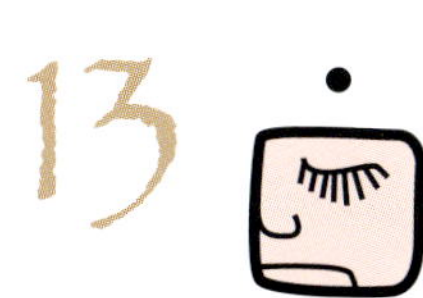 June 11 2008 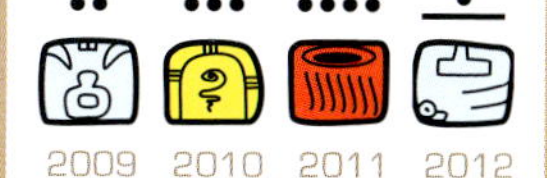 2009 2010 2011 2012	Gemini 14 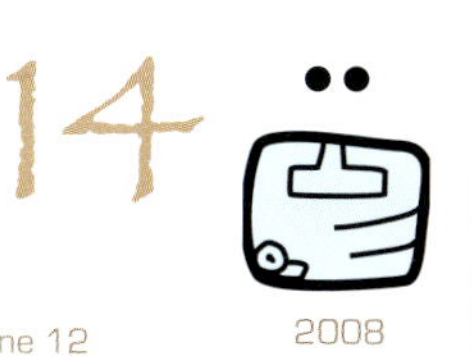 June 12 2008 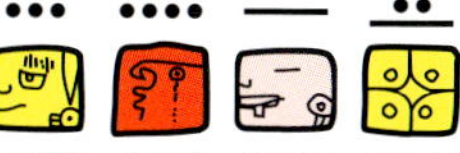 2009 2010 2011 2012
Gemini 15 June 13 2008 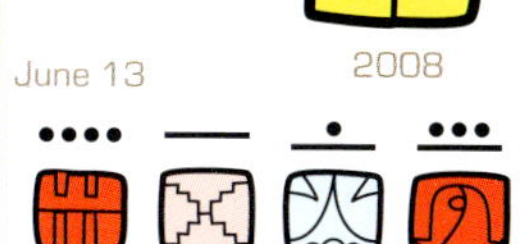 2009 2010 2011 2012	Gemini 16 June 14 2008 2009 2010 2011 2012	Gemini 17 June 15 2008 2009 2010 2011 2012	Gemini 18 June 16 2008 2009 2010 2011 2012	Gemini 19 June 17 2008 2009 2010 2011 2012	Gemini 20 June 18 2008 2009 2010 2011 2012	Gemini 21 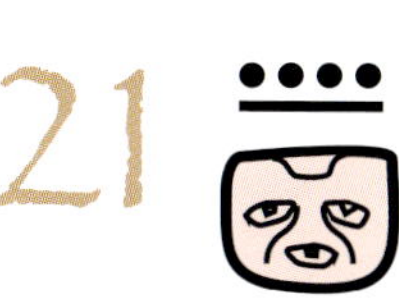 June 19 2008 2009 2010 2011 2012
Gemini 22 June 20 2008 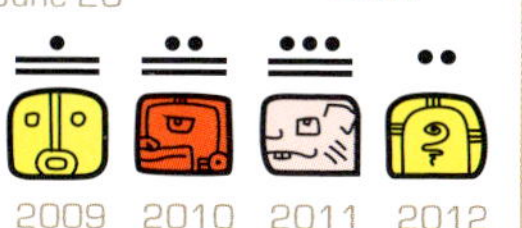 2009 2010 2011 2012	Gemini 23 June 21 2008 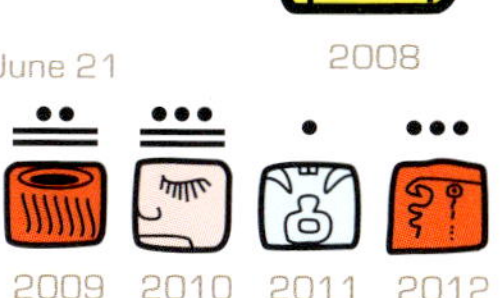 2009 2010 2011 2012	Gemini 24 June 22 2008 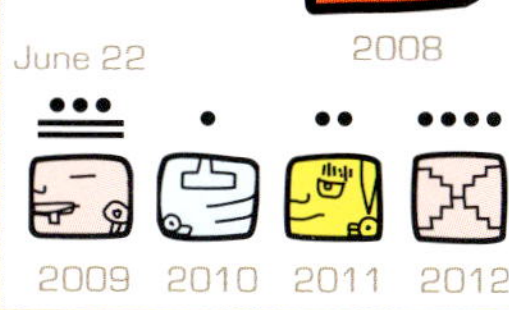 2009 2010 2011 2012	Gemini 25 June 23 2008 2009 2010 2011 2012	Gemini 26 June 24 2008 2009 2010 2011 2012	Gemini 27 June 25 2008 2009 2010 2011 2012	Gemini 28 June 26 2008 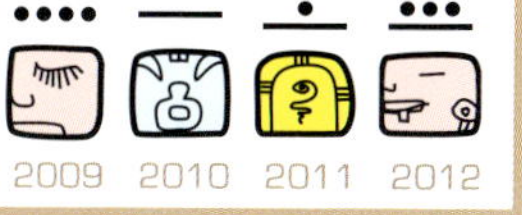 2009 2010 2011 2012

MOON

Cancer

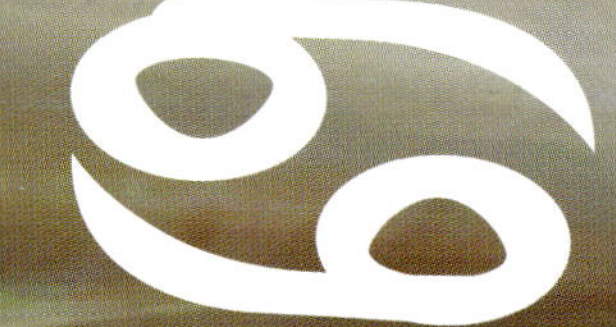

Cancer 1	Cancer 2	Cancer 3	Cancer 4	Cancer 5	Cancer 6	Cancer 7
June 27 2008	June 28 2008	June 29 2008	June 30 2008	July 1 2008	July 2 2008	July 3 2008
2009 2010 2011 2012	2009 2010 2011 2012	2009 2010 2011 2012	2009 2010 2011 2012	2009 2010 2011 2012	2009 2010 2011 2012	2009 2010 2011 2012

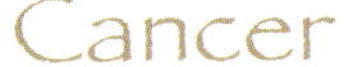
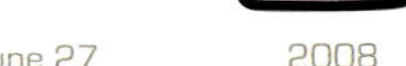
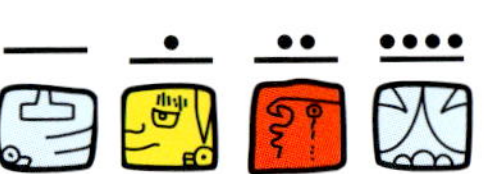
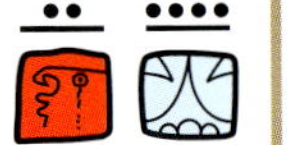
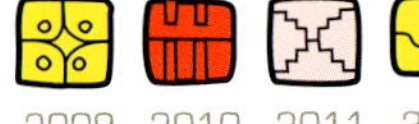
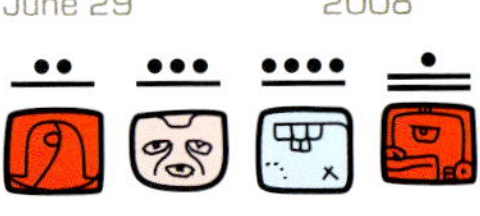
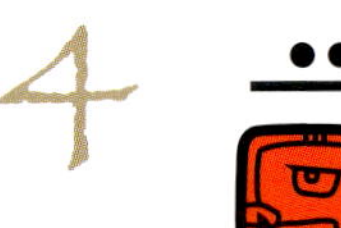
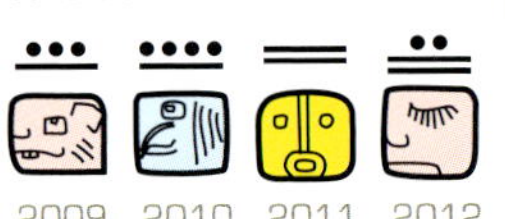
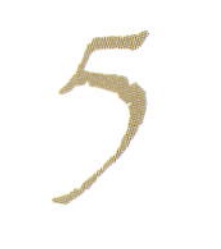

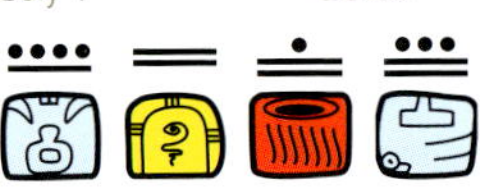
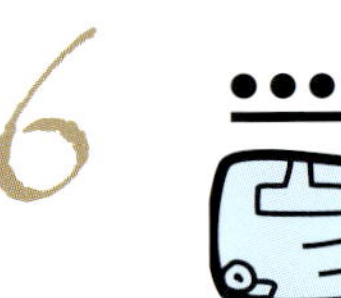
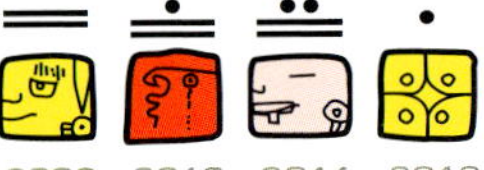

Cancer 8	Cancer 9	Cancer 10	Cancer 11	Cancer 12	Cancer 13	Cancer 14
July 4 2008	July 5 2008	July 6 2008	July 7 2008	July 8 2008	July 9 2008	July 10 2008
2009 2010 2011 2012	2009 2010 2011 2012	2009 2010 2011 2012	2009 2010 2011 2012	2009 2010 2011 2012	2009 2010 2011 2012	2009 2010 2011 2012

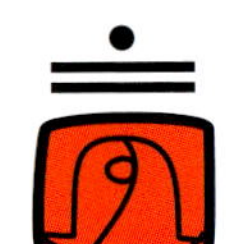
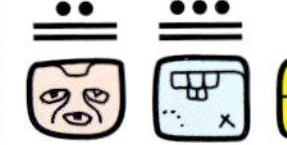
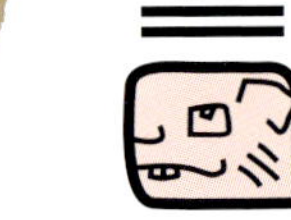
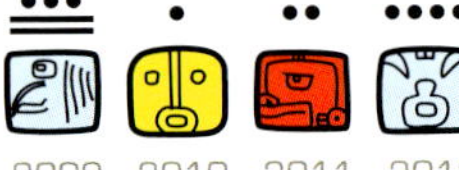
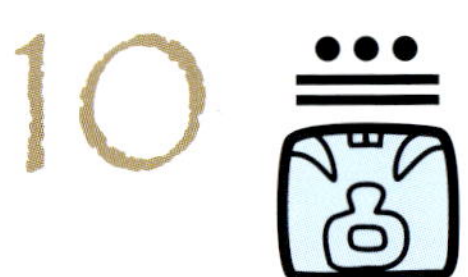
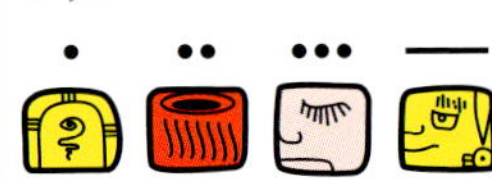
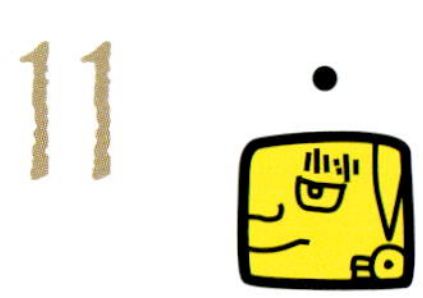
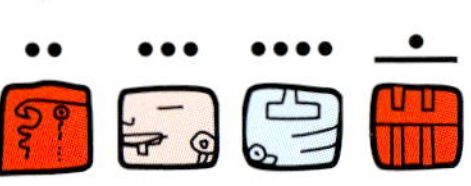
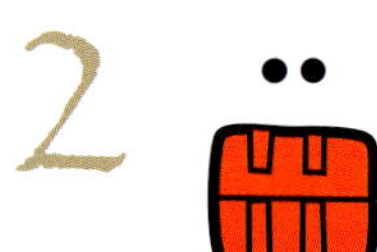
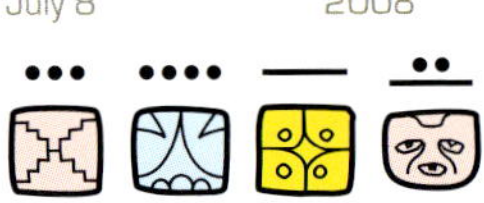
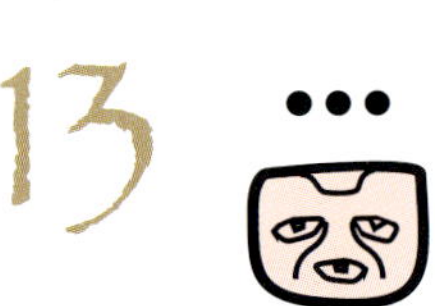
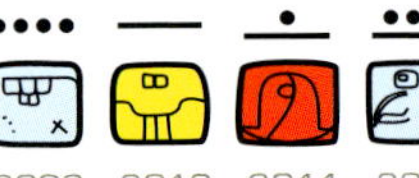
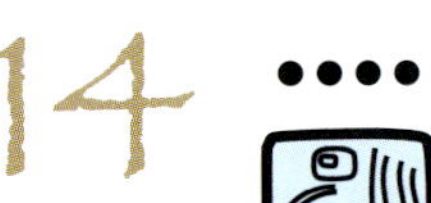
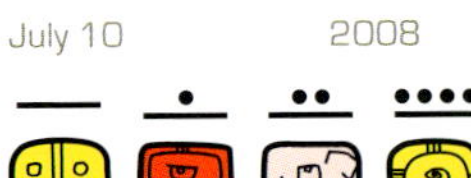

Cancer 15	Cancer 16	Cancer 17	Cancer 18	Cancer 19	Cancer 20	Cancer 21
July 11 2008	July 12 2008	July 13 2008	July 14 2008	July 15 2008	July 16 2008	July 17 2008
2009 2010 2011 2012	2009 2010 2011 2012	2009 2010 2011 2012	2009 2010 2011 2012	2009 2010 2011 2012	2009 2010 2011 2012	2009 2010 2011 2012

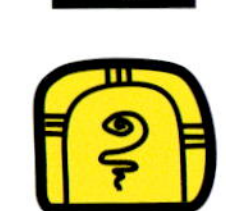
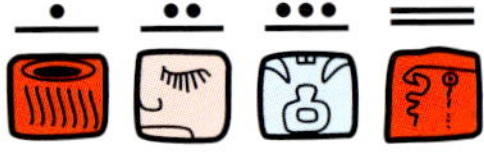
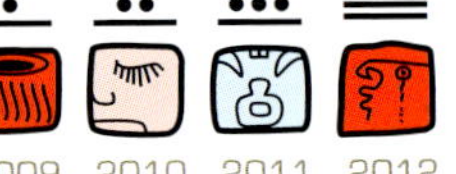
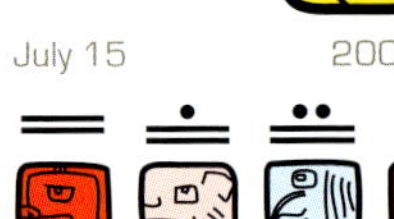
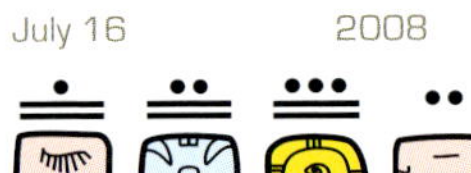

Cancer 22	Cancer 23	Cancer 24	Cancer 25	Cancer 26	Cancer 27	Cancer 28
July 18 2008	July 19 2008	July 20 2008	July 21 2008	July 22 2008	July 23 2008	July 24 2008
2009 2010 2011 2012	2009 2010 2011 2012	2009 2010 2011 2012	2009 2010 2011 2012	2009 2010 2011 2012	2009 2010 2011 2012	2009 2010 2011 2012

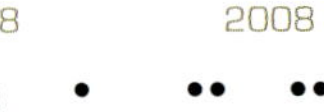
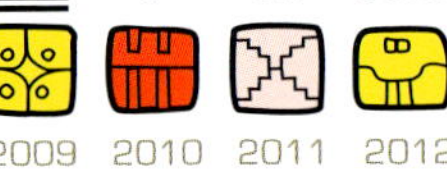
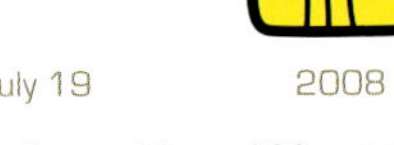
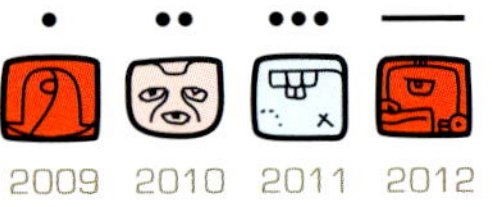
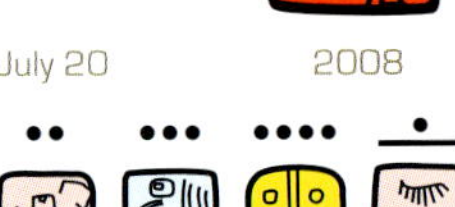
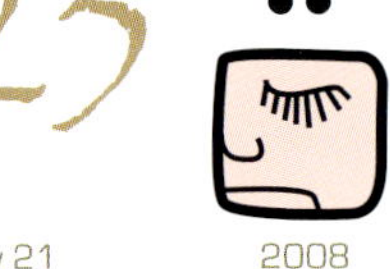
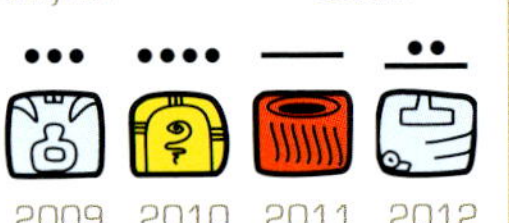
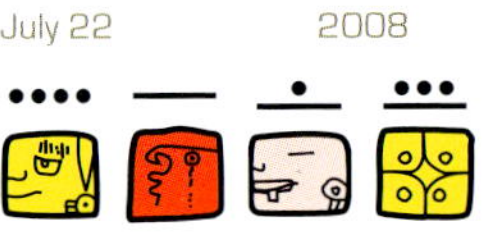
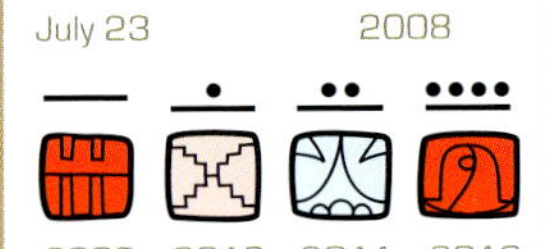
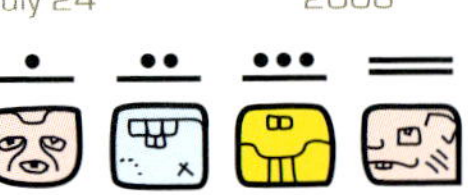

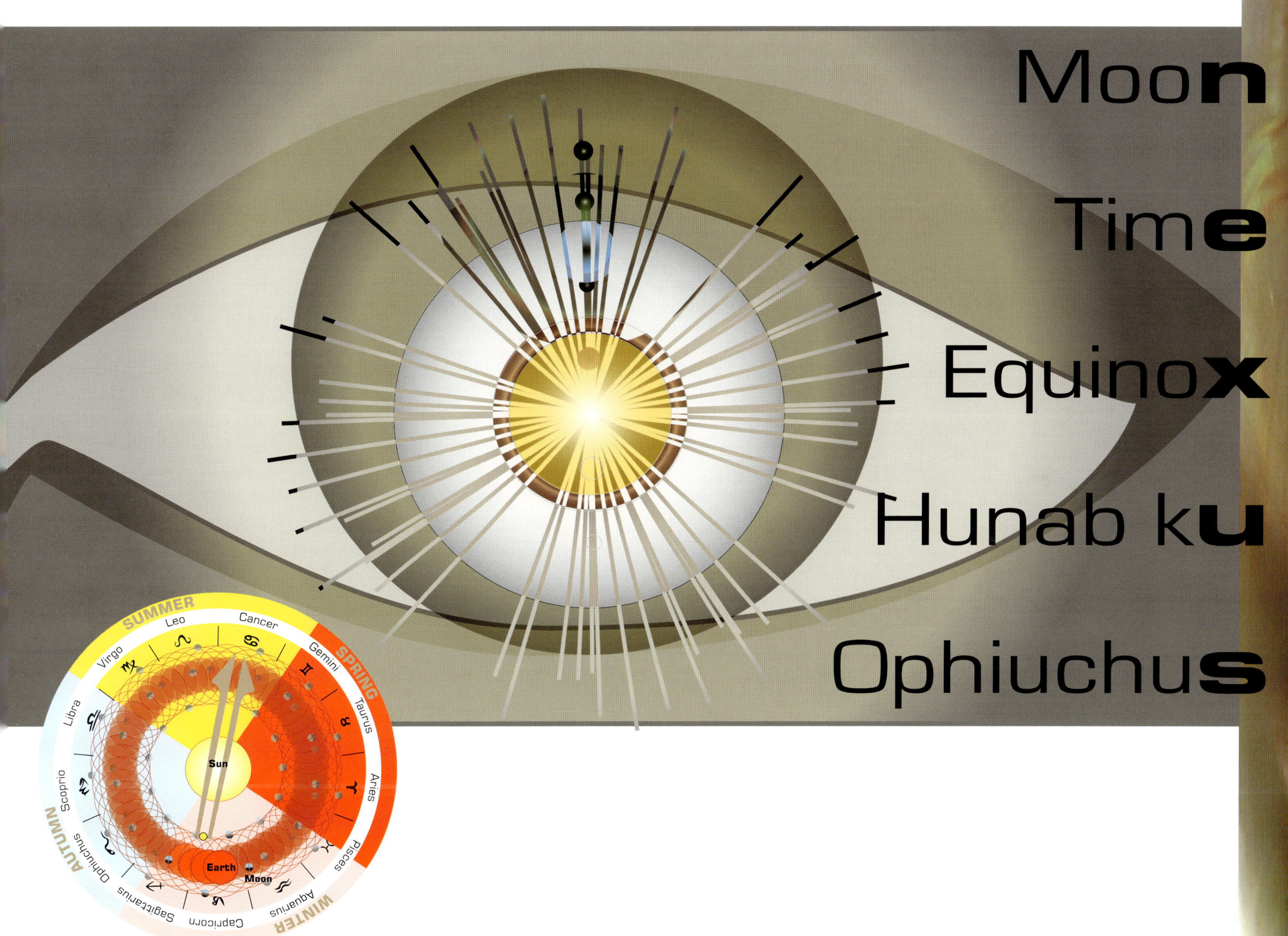
Moon
Time
Equinox
Hunab ku
Ophiuchus
SUMMER
Leo
Cancer
Virgo
Gemini
SPRING
Taurus
Libra
Aries
Sun
Scoprio
AUTUMN
Pisces
Ophiuchus
Earth
Moon
Sagittarius
Capricorn
Aquarius
WINTER

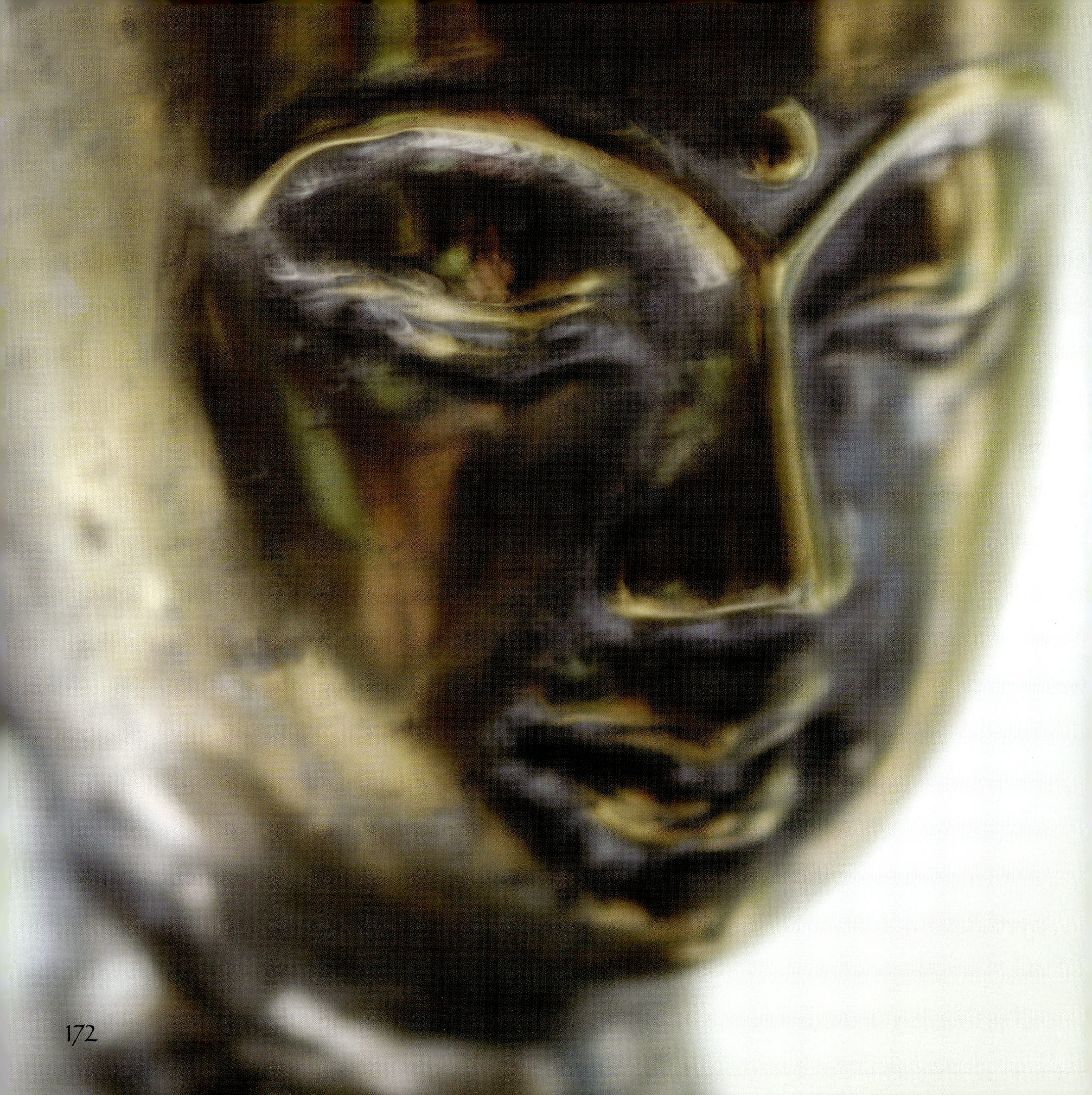

A DATELESS DAY To honor Earth

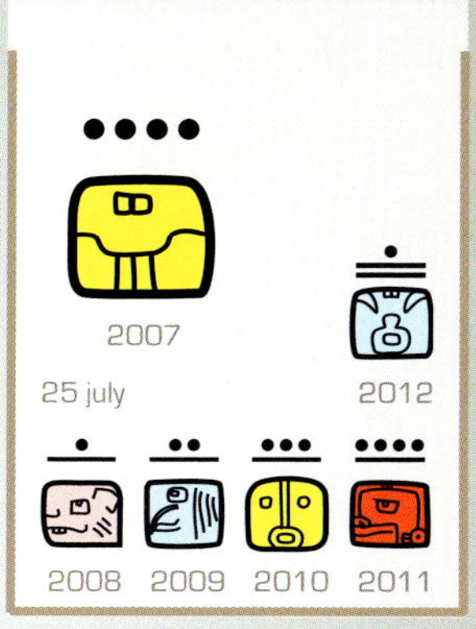

Dateless-day in honor of Gaia

A year of 13 months of 28 days gives 364 days, 1 day short, July 25. This day has no date, all free to think about Earth and other life. The Zolkin count goes on for 2,500 years, so also at the July 25.

Leap year

Every 4 years an extra Dateless-day.

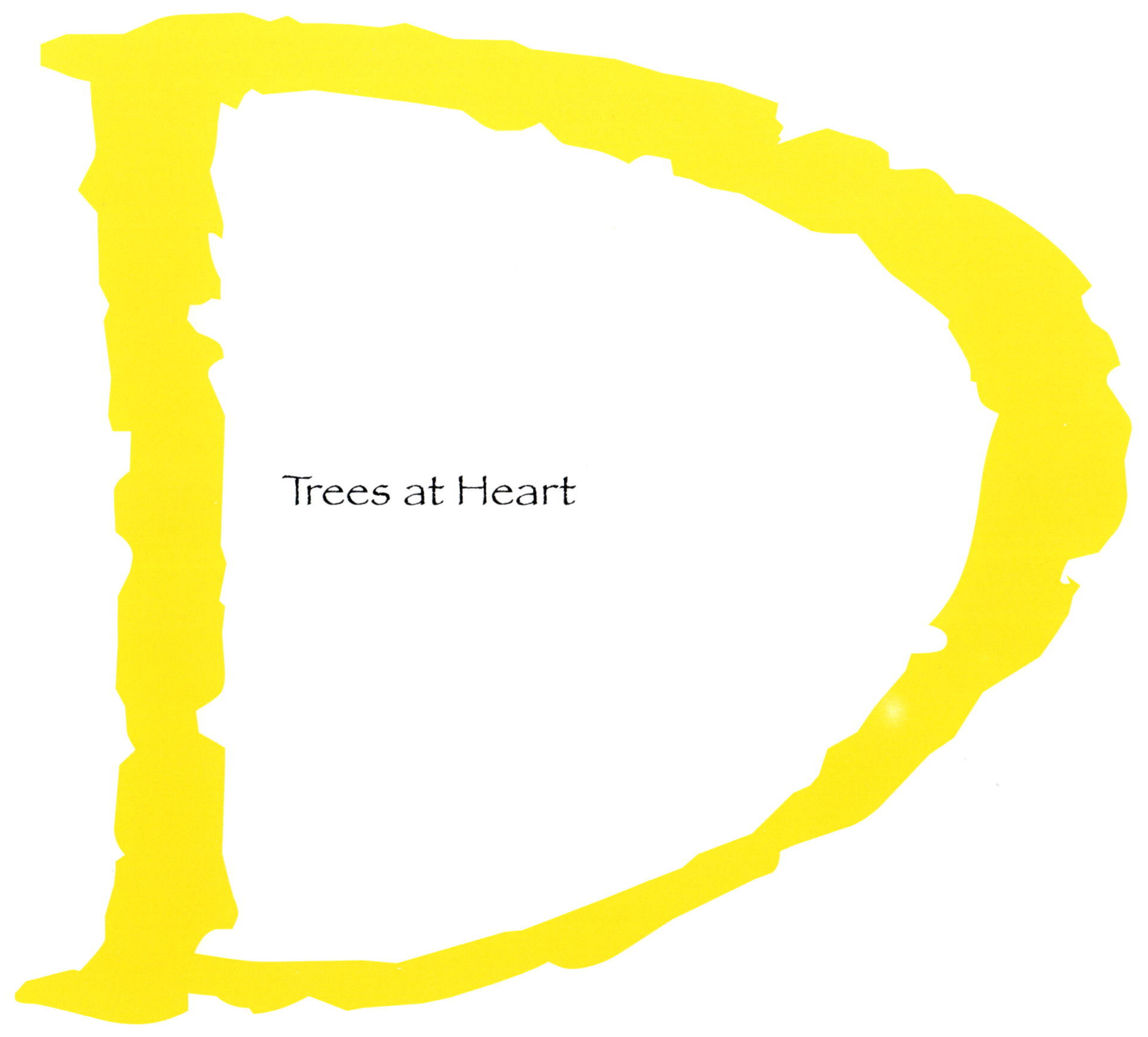

Trees at Heart

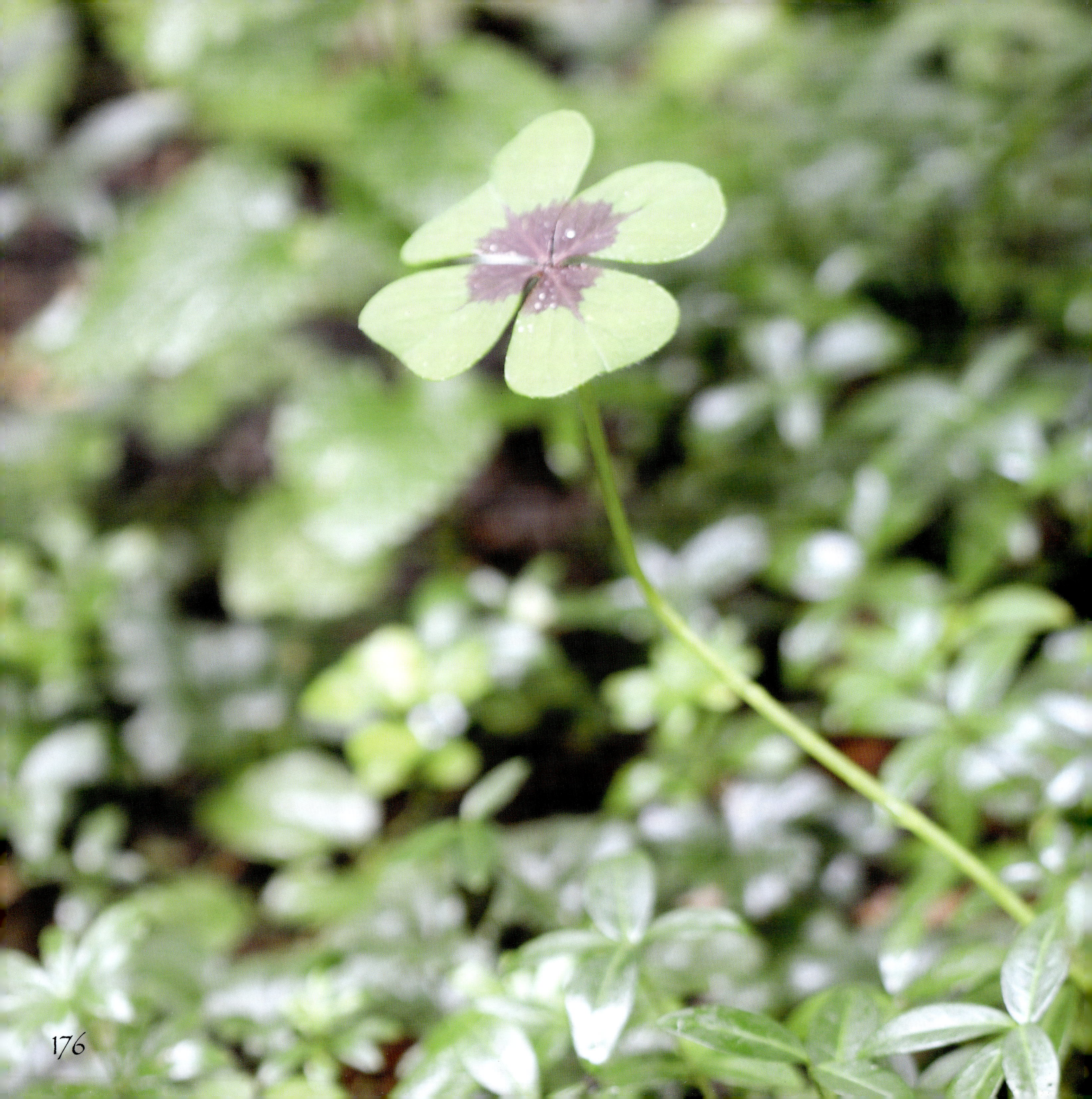

trees at heart

Trees at Heart

For children by children.

Make a painting, sell it, buy a tree and plant a whole forest together.

Trees for the future, for lots of animals to live.

Empower children, they can make a difference all by themselves.

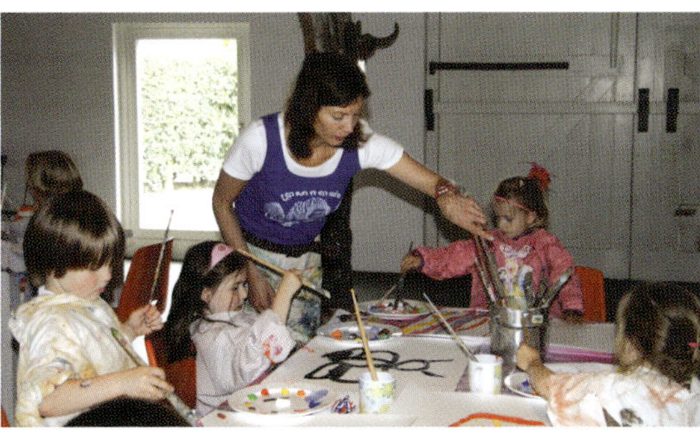

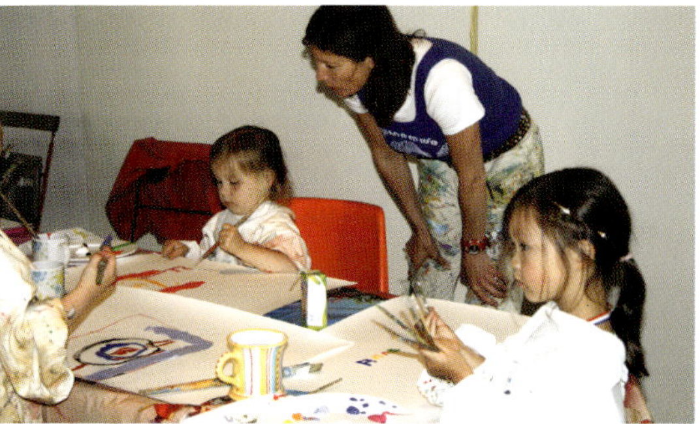

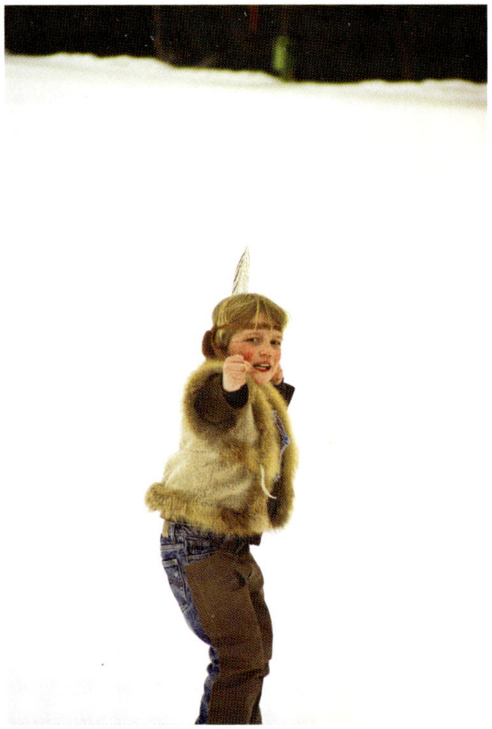

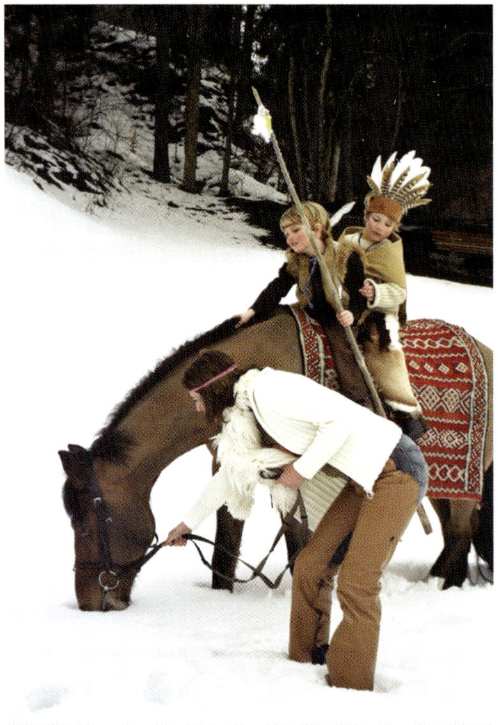

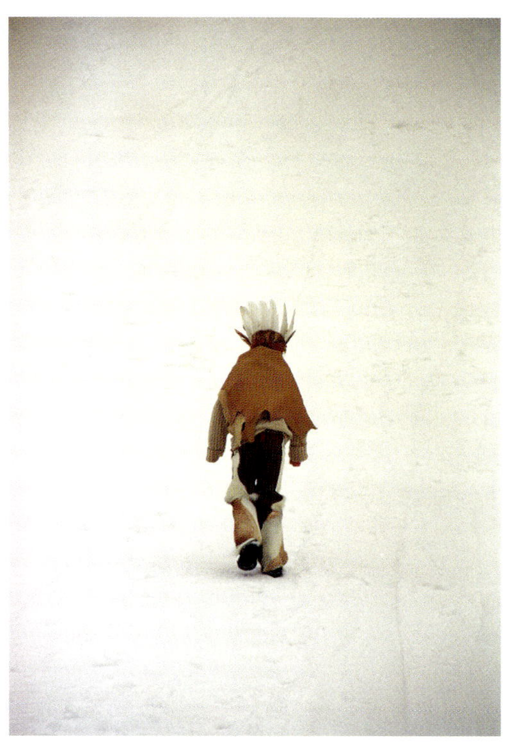

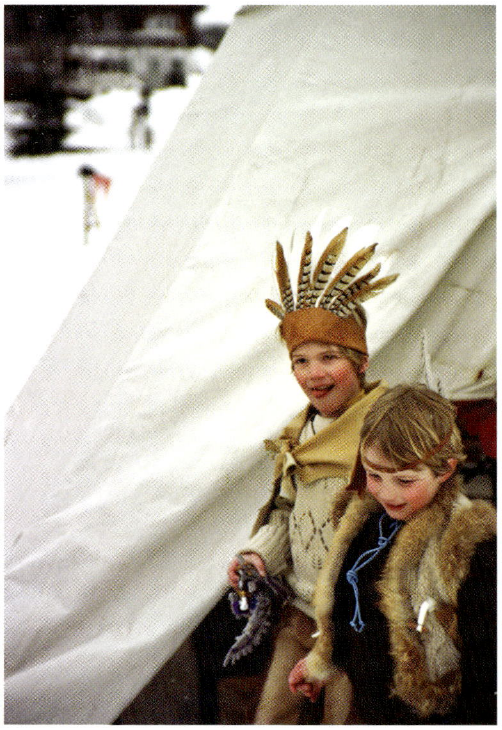

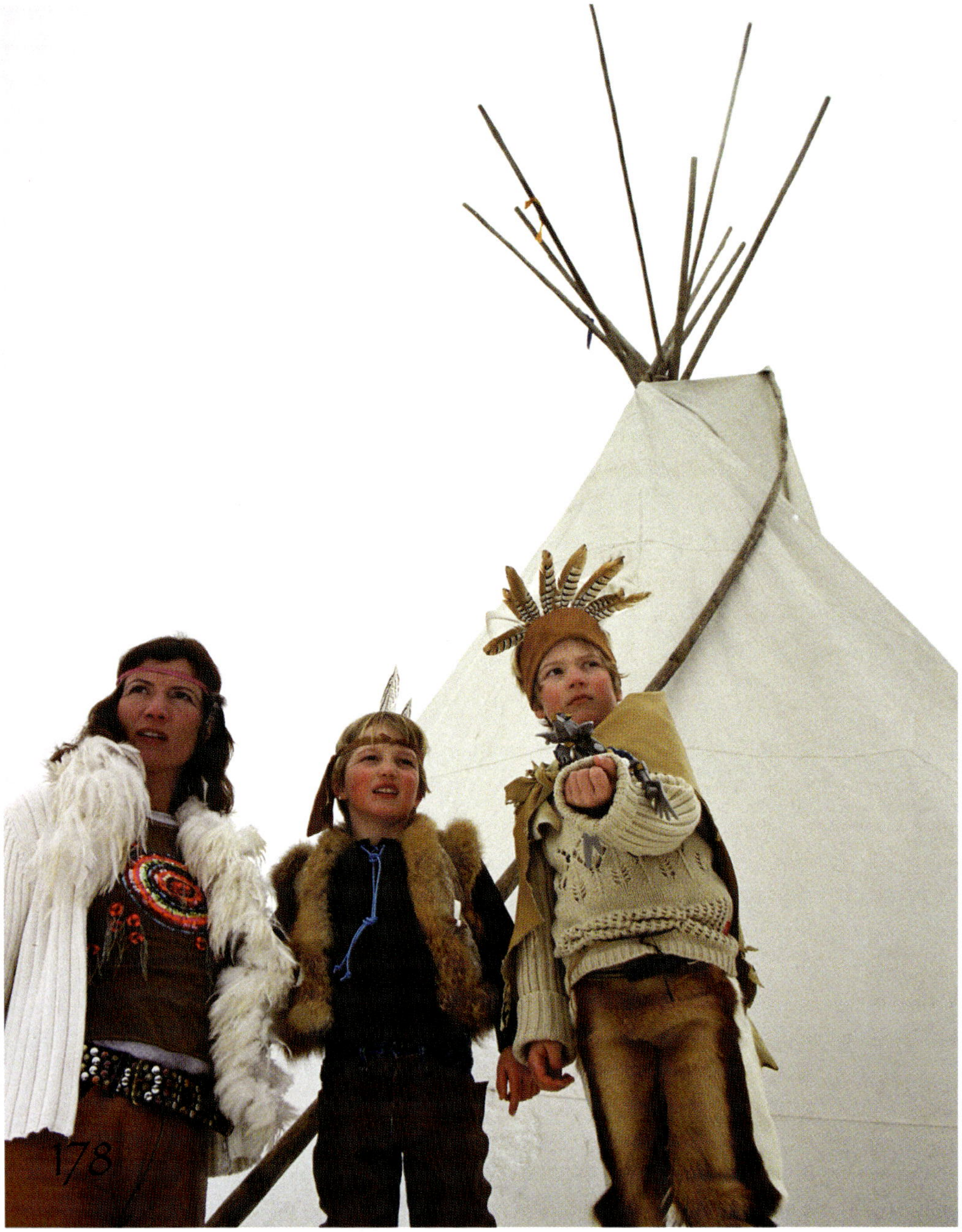

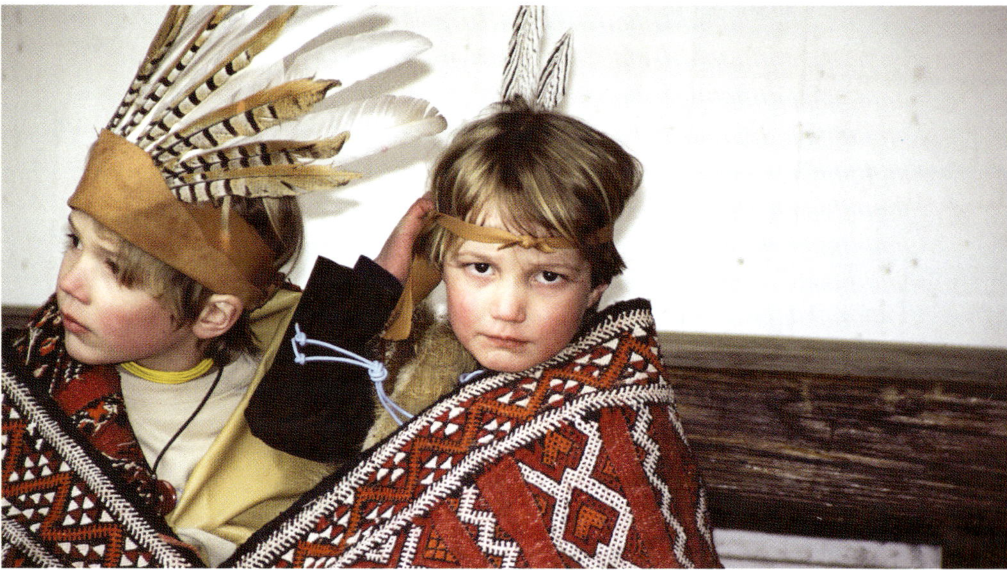

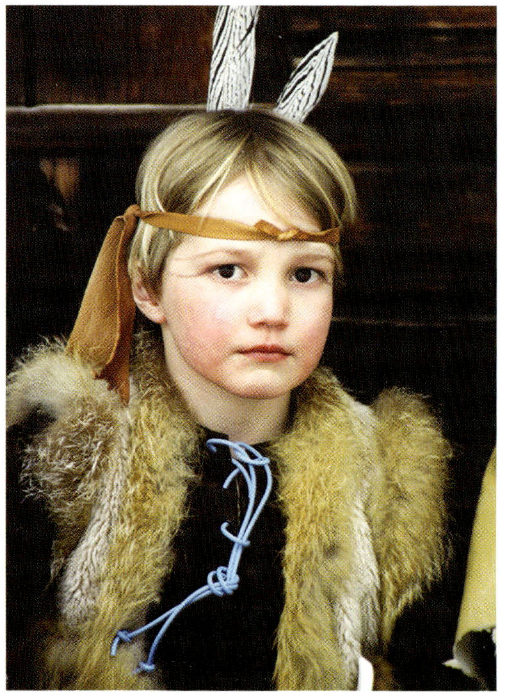

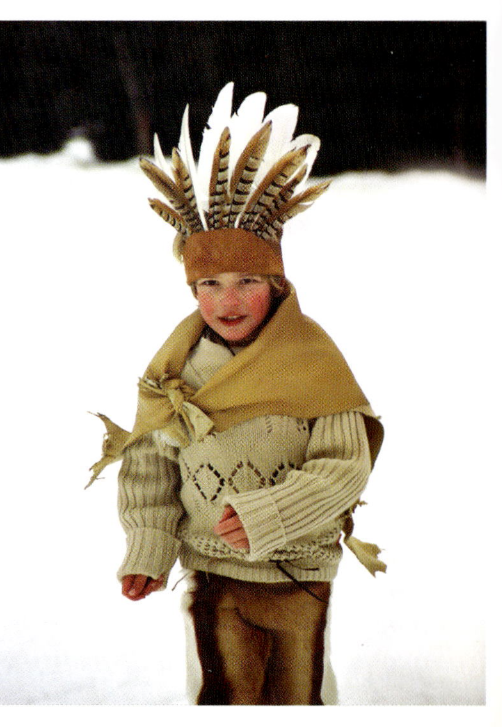

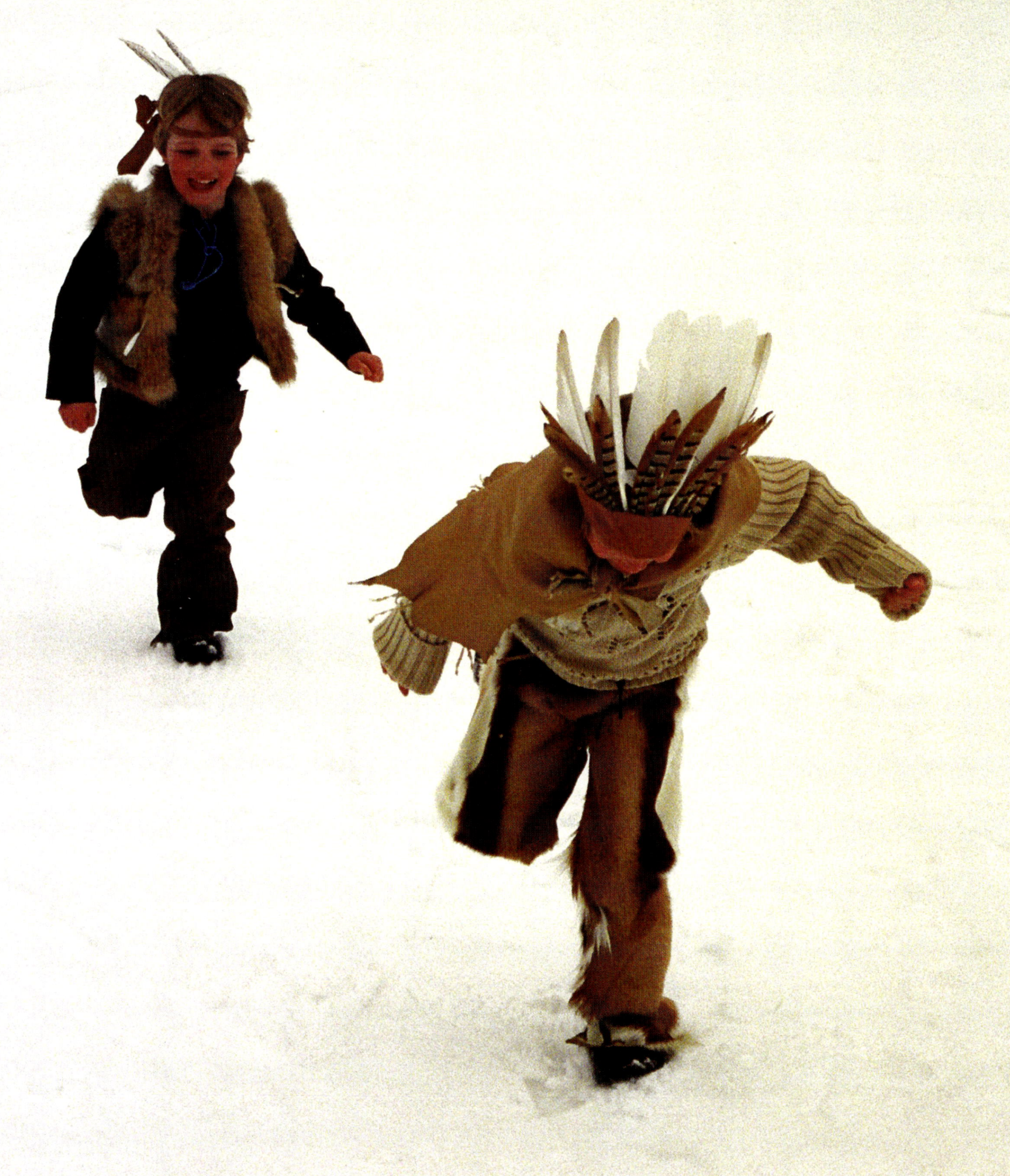

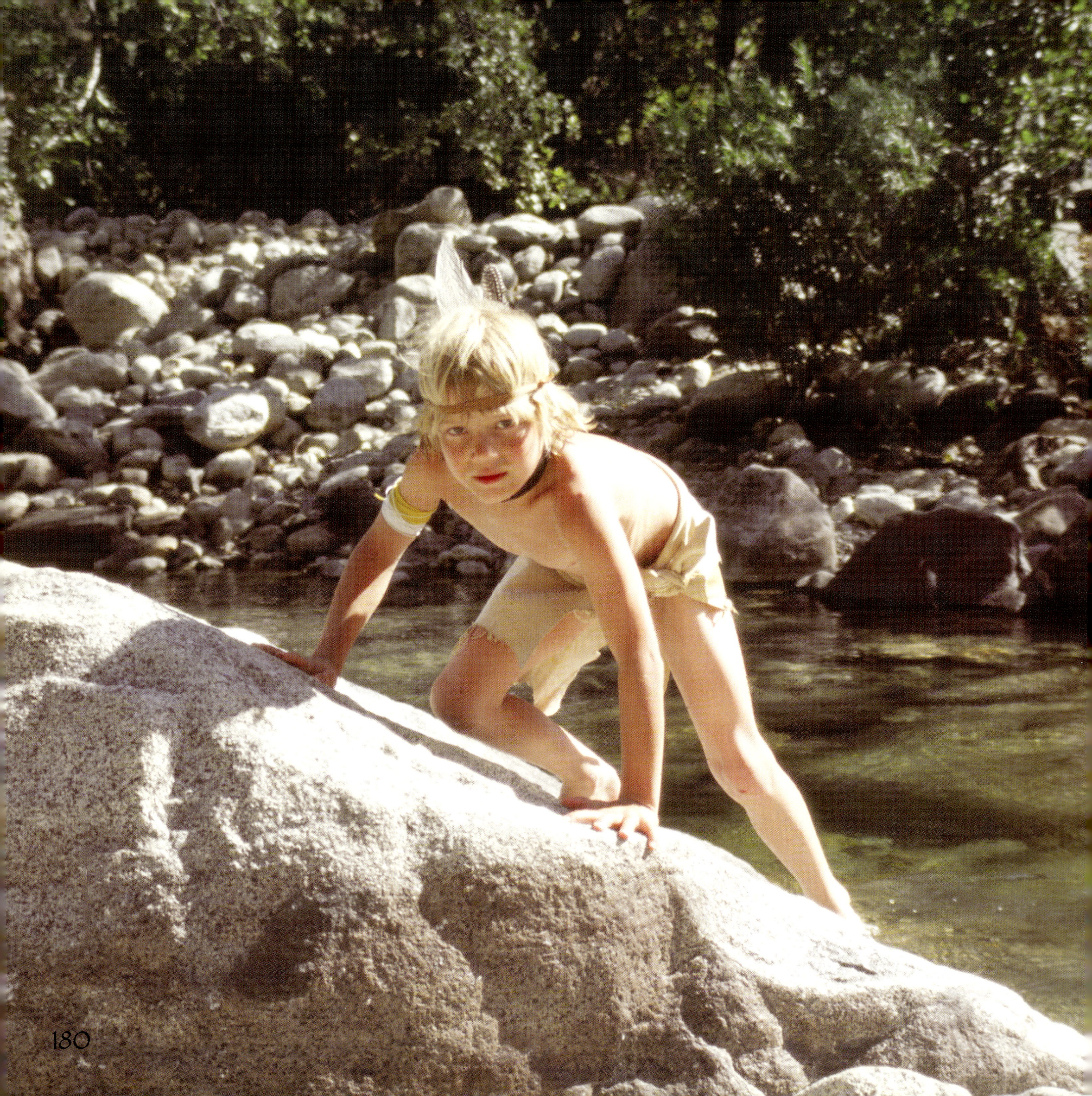

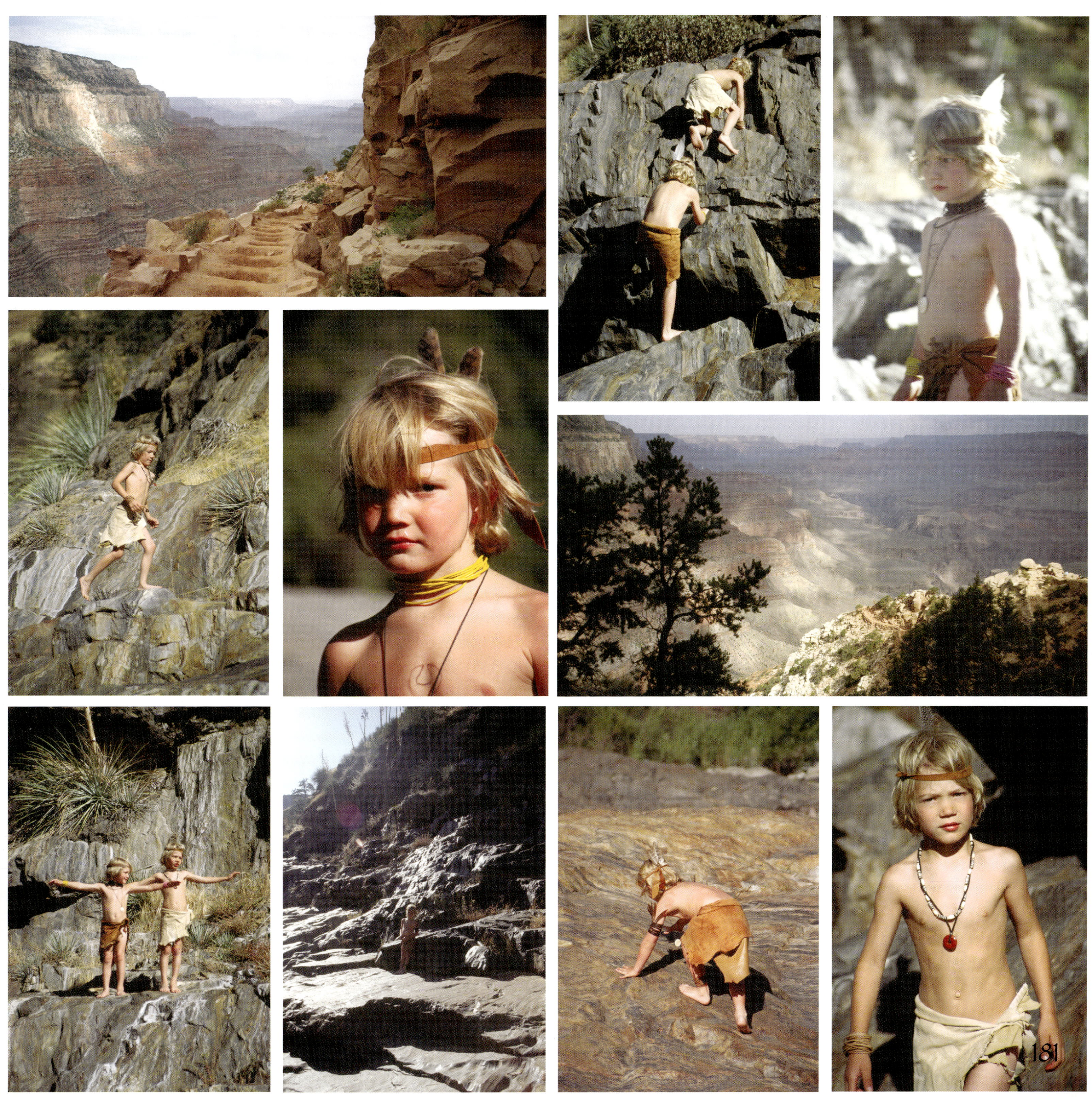

Thanks To

Maximiliaan & Titus: “Mom, you have such cool ideas.” Thanks for your obsession with the Egyptian museum in Leiden and for both your boundless interest and inspiring ideas, because of the two of you this book has a fact. Werner Paul for your never failing trust and encouragement and for all the freedom I have in my life because of you! Marjan Bosman, my oldest friend, always good for a laugh, for your good advice, your interest in my art (not so much in what’s behind it!). Anouk Jutta Heidweiler, my friend from the academy, too vague to follow then, a sublime artist, a beautiful person and a pure soul who thinks I am the vaguest of all. Sally Collins & Simon Bennie for complementing me on my English: “Why don’t you write in English so I can read it.” Here it is Sally, something to chew on! Thanks for introducing me to spirituality. Maya Wildevuur for encouraging me to continue painting and pursuing combination of arts. Marianne Geytenbeek for advising and conversing on astrology and the Maya Calendar. Riet de Vos for everything, for being so exceptional and so basic, a rare combination. Peter Schenk for the inspiring painting group sessions, spent in your atelier, in a good atmosphere. Eric Fabery de Jonge for helping with the jewellery in such a relaxed, always joyful way. Hein Koningsveld, for letting me start my graphic design ‘business’ within your company. Renée Graadt van Rogge for your moral understanding and support. E. Vogel for the sublime craftsmanship in drum making. Ageeth de Jong for your enthusiasm. Monique Rutter for persistent interest, showing me different angles. Brigitte Pompe for kicking me into action: “Get it done, so we all know what it is about.” Mariska de Blanken, for rescuing and teaching me how to put this whole project into a book.

Coert and Ernst van der Burg, my brothers, for always being interested in my art. Last but not least, Ingeborg van der Burg, my elder sister, for all the advice, your enthusiasm, and endless patience in correcting my dyslexic spelling. And for diving into these spiritual matters, not your thing, and the wonderful conversations about 2012. I am so very sad you are no longer here for the result, nor for the publishing of your own book. We were so looking forward to publishing our first book both in the same year.

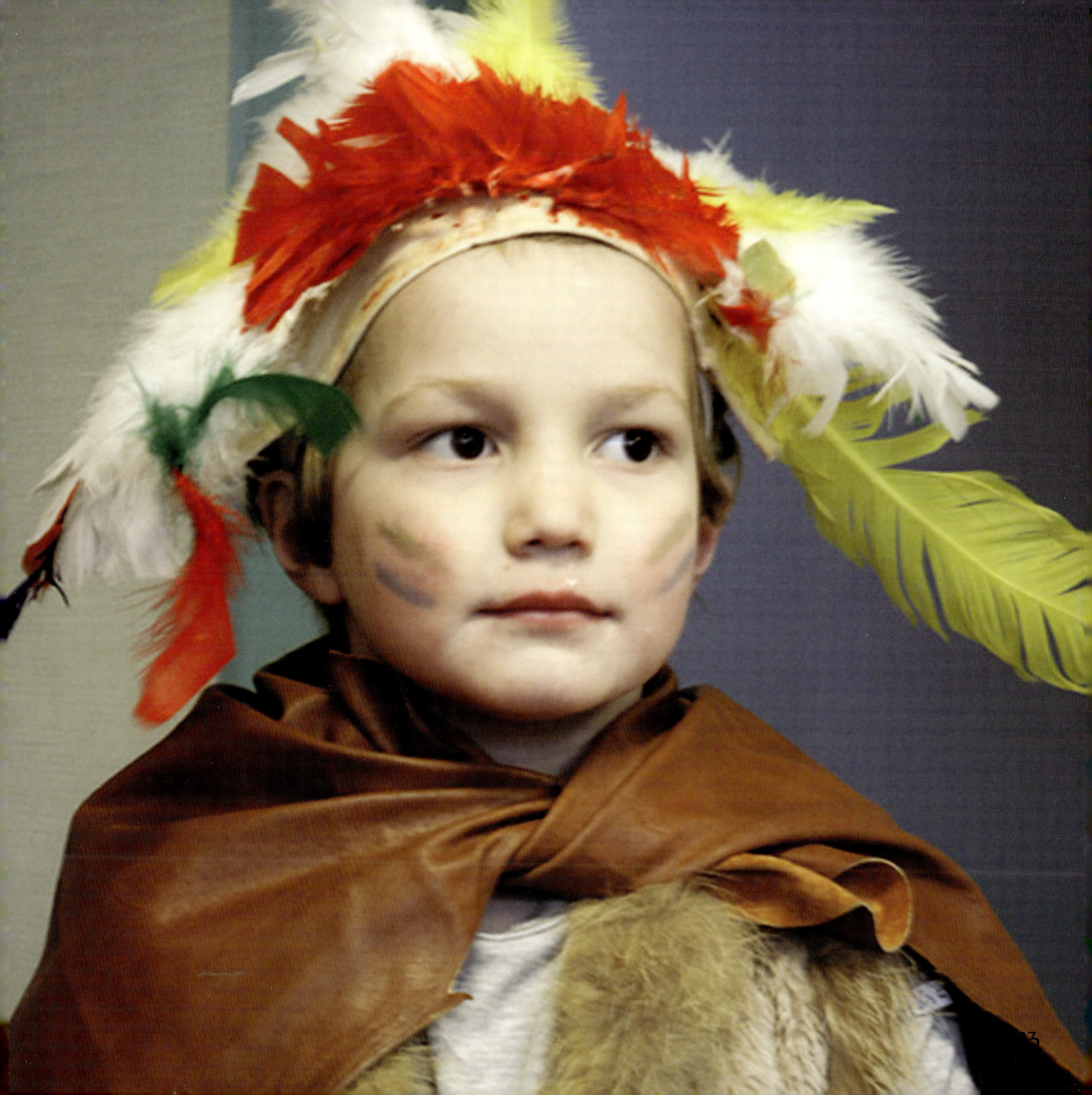

Literature

~ Maya cosmogenesis 2012 by John Mayor Jenkins, published by Bear & Co, Rochester, Vermont

~ Galactic Alignment, the transformation of consciousness according to Mayan, Egyptian and verdic Traditions by John Major Jenkins, published by Bear & Co, Rochester, Vermont

~ The Mayan Factor by José Argüelles, published by Bear & Co, Rochester, Vermont

~ Time & Technospere, by José Argüelles, published by Bear & Co, Rochester, Vermont

~ The Mayan Calendar and the transformation of consciousness, by Carl Johan Calleman, Ph.D, published by Bear & Co

~ The Mayan Oracle, return path to the stars, by Ariel Spilsbury & Michael Bryner, published by Bear & Co

~ Wat wisten de Maya's? Peter Tonen

~ A Way to Live the Ancient Maya Calendar, 13-moon diary of natural time, by Nicole E. Zonderhuis and Sylvia Carrilho, Frontier Publishing, Amsterdam & Kempton, USA

~ Reading the Maya glyphs, by Michael D. Coe and Mark Stone, published by Thames and Hudson, London

~ Mayan Astrology, by Aluna Joy Yaxk'in, published by Center of the Sun, Sedona

~ Anastasia, by Vladimir Megré, published by Ringing Cedars Press, Hawaii, Usa

~ Cosmic Vision, the dawn of the integral theory of everything, by Ervin Laszlo, (Kosmische visie) published by Ankh-hermes

~ You can change the world - the global citizen's handbook for living on planet Earth, by Ervin Laszlo, published by SelectBooks

~ The ancient secret of the flower of life I and II, by Druvalo Melchizedek,

~ Nothing in this book is true, but it's exactly how things are, by Bob Frissell, published by Frog Ltd., North Atlantic Books, California

~ The Pleiadian Agenda, a new cosmology for the Age of Light, by Barbara Hand Clow, published by Bear & Co, Santa Fe, USA

~ Alchemy of the nine dimensions, Decoding the vertical Axis, Crop Circles, and the Maya Calendar, by Barbara Hand Clow, Bear & Co

~ 2012, Der Aufstieg der Erde in die fünfte dimension, Ute Kretzschmar, Falk-Verslag, Seeon, Germany

~ Illustrated A-Z of Understanding Star Signs, by Kim Farnell, Flame Tree Publishing, London

~ Egyptian Cosmology, by Mustafa Gadalla, published by Bastet Publishing, Pennsylvania

~ The Invisible Universe, by David Malin, published by Callaway Editions, New York

~ Celestial Gallery, by Romio Shretha, published by Callaway New York

~ The 12th Planet, by Zecharia Sitchin, published by Frontier Publishing, Amsterdam

~ A Brief History of Time, by Stephan W. Hawking, Bantam Press, London

~ Vom richtigen Zeitpunkt das Mondjahr, by Johanna Paungger & Thomas Poppe, Hugendubel Verslag, München

~ How can you sell the air?, by chief Seattle of Suquamish 1854, The Book Publishing Company, Summertown, USA

~ Medicine Cards, the discovery of power through the ways of animals, by Jamie Sams & David Carson, publisher: Bear & Co

~ Seven Arrows, by Hyemeyohsts Storm, published by Ballantine Books, New York

~ The Medicine Wheel, Earth Astrology, by Sun Bear & Wabun, published by Prentice-Hall Inc., New Jersey

~ North American Indian Jewelry and Adornment, by Lois Sherr Dubin, Abrams Publishers New York

~ Khmer mythology, secrets of Angkor by Vittorio Roveda, published River Books Co, Bangkok

~ In an Eastern Rosegarden, by Inayat Khan, London 1920

~ Sacred Architecture, Explore and Understand Sacred Spaces, by Caroline Humphrey & Piers Vitebsky

~ The Healing Energy of Trees, by Partice Bouchardon, published by Gaia Books, London

~ The hypnotic Power of Crop Circles, by Bert Janssen published by Frontier Publishing, Amsterdam & Kempton, USA

~ The Silent Takeover, Global Capitalism and the Death of Democracy, by Noreena Hertz, published by Harper Business, New York

From the same publisher:

13-Moon Diary of Natural Time 2007 – 2008 *A Way to Live the Ancient Maya Calendar*
Nicole E. Zonderhuis & Sylvia Carrilho

More and more people know about the Mayan prophecy and their unique way to deal with Time. The Mayan Tzolkin Calendar of 260 days marks each day with its own special energy. It is wise not to make important decisions on a 'Moon day' when emotions dominate. On a 'Wizard day' (dominated by timelessness) hours can get lost, so try to avoid tight schedules. Also, each month/Moon has its own energy and totem animal.
With this 13-Moon Diary you can see at a glance what day it is, both according to the 'ordinary' 12-month Gregorian calendar as well as in the 13-Moon Calendar. You can find out its meaning and plan your activities accordingly. Or you can check the date afterwards to see if it made sense. You can also learn to calculate your own personal 'Mayan horoscope', and those of your friends.
This desk diary is a combination of the Gregorian calendar and the 13-Moon Calendar, or Dreamspell Calendar by José Argüelles. We have tried deliberately to keep this diary clear, in order to avoid excessive information. This means this diary is not complete. The Divine Being of Time is very complex; many books have been written about it. For more detailed information, we would like to refer to those books and to the various websites dedicated to the 13-Moon Calendar. An overview has been included. For practical reasons, we have decided to work with the week schedule as we know it from the 'ordinary diary', starting off on Monday, the first working day.
This full color day-planner offers many pages of clear explanation of the 13-Moon Calendar (Wavespells, Solar Seals, Tones) illustrated with lots of images. It has ofcourse all the features of a contemporary diary. The diary starts on July 26 2007 and covers the whole White Lunar Wizard Year (till July 25, 2008)
272 Pages. Hardcover. GBP 13.20 / USD 20.13

Maya Pocket Planner by Nicole E. Zonderhuis & Sylvia Carrilho

Our month has everything to do with the Moon. The time between two Full Moons once determined our month. The Moon orbits our Earth precisely 13 times, when we orbit the Sun just once. You could see this very clearly if you would look down on our solar system. That's why the 13-Moon Calendar consists of 13 months/Moons of each 28 days, 13 x 28 = 364 days.
The 365th day is called the Day Out of Time. It is the day between the years. It is a holiday of connections; the old is released and the intention for the New Year is set. Each year the Day out of Time is celebrated on July 25.
The ancient Maya and Egyptians started their year on July 26. They were aware of the fact that on July 26, the Sun and Sirius both arise at our horizon at exactly the same moment and exactly the same place. No other synchronicity between those 3 heavenly bodies occurs throughout the year. That's why the 13-Moon Calendar starts on July 26.
In the 13-Moon Calendar the year is called after the Tone and Seal of the first day of the year. We are now in the White Lunar Wizard year. At July 26, 2008 starts the Blue Electric Storm Year. This pocket planner is meant to help you integrate the Tzolkin in your everyday life.
144 Pages. Small booklet. GBP 6.90 / USD 9.99

Egypt: Image of Heaven by Wim Zitman

The ancient Egyptians were the first geographical planners to develop a system in order to establish an 'image of heaven' on earth. According to the astronomer Moore, "the precision [of the Egyptians] was amazing by any standards, and there is no doubt that the Pyramids were astronomically aligned". This book completes Zitman's ten year research into how the Pyramid Field depicts The Constellation of Horus, the deity who bore the meaning of power and invincibility – and who guarded the Pharaoh. Rather than randomly pick certain pyramids, Zitman is the first who has been able to make sense of the entire era of pyramid building. Is this depiction of Heaven on Earth an inheritance of the mythical Followers of Horus, who were said to rule Egyptian in Predynastic times? In this popularly scientific book, Zitman reveals how time and space were perceived by the Egyptians as sacred ingredients, which they mixed into a divine master plan, which for the first time will be unveiled in its entirety.

311 Pages. Paperback. 12.99 GBP / 19.90 USD

Land of the Gods ~ *How a Scottish landscape was transformed to become Arthur's "Camelot" by* Philip Coppens

Land of the Gods is the story of the ancient inhabitants of the Lothians and the Borders, whose accomplishments are visible in Cairnpapple, Traprain Law and other ancient monuments. They accentuated the region's unique volcanic landscape to make it reflect their mythology, which spoke of gods descending from Earth and the local ruler being a representative of the sun god – Loth.

Throughout history, the land remained special: the Romans did not conquer the Gododdin, as they called the inhabitants. And when the Romans were retreating from Britain and for the first time, neighbouring tribes tried to lay claim to the land, a magnificent warrior appeared, who would fight for the survival of his land. He was remembered as Arthur and his Camelot was the Lothians and Borders region.

Unfortunately, after his reign, the region was overrun and his tribe, the Gododdin, fled to Wales, where they would speak of their magical kingdom and the mythical hero… The legend of Arthur was born… but the history of Camelot forgotten.

252 Pages. Paperback. 13.90 GBP / 17.95 USD

Saunière's model and the Secret of Rennes-le-Château *The priest's final legacy that unveils the location of his terrifying discovery* André Douzet

In 1916, Berenger Saunière, the enigmatic priest of the French village of Rennes-le-Château, created his ultimate clue: he went to great expense to create a model of a region said to be the Calvary Mount, indicating the "Tomb of Jesus". But the region on the model does not resemble the actual lay-out of Jerusalem. Did Saunière leave a clue as to the true location of his treasure? And what is that treasure?

After years of research, André Douzet discovered this model, never collected from the model maker by Saunière, who had died just before the model's completion. Backed by evidence showing correspondence between Saunière and the model maker, Douzet also reveals much new evidence, including the revelation that Saunière spent large amounts of time and money in the city of Lyons, often on exotic and high tech photographic instruments. And for the first time, it is shown Saunière met some very interesting people from esoteric, in particular Martinist, circles in Lyons. This body of evidence for the first time demonstrates there is indeed a true mystery surrounding this village priest – a theory widely speculated on so far by others authors, but seldom if ever backed by evidence.

The model is the only real clue Saunière left behind as to the nature and location of his treasure – and is unveiled in this book, which includes pictures and detailed drawings of the model, among many other things. It also reveals the location of the region where Saunière had located his treasure… and where Douzet himself has so far recovered large quantities of precious and semi-precious materials. Above all, Douzet demonstrates that grounds of Perillos not so much hold a "treasure" (though present), but rather a "secret", and that this secret was the true importance of Saunière's mystery; a secret that is said to be of vital importance – and terrifying force.

116 Pages. Paperback. GBP £ 7,99. USD 12.00.

The Secret Vault *The secret societies' manipulation of Saunière and the secret sanctuary of Notre-Dame-de-Marceille*

André Douzet & Philip Coppens

Was Berenger Saunière, the priest at the centre of the enigma of Rennes-le-Château, controlled by a secret society? Yes is the answer, but which one? Freemasonry? The Rosicrucians? None of the above; the group was known as the A.A., said to be the successors of the Compagnie de St Sacrement, a 17th century secret society that was prohibited by the French king Louis XIV. Both were said to protect "the secret", of such gigantic proportions that its nature has never been revealed; no member was able to divulge membership or what "the secret" was – and that is exactly what Saunière did when his superiors questioned him late in life: he remained silent.

In this brilliant example of investigative research, the authors have been able to uncover a hidden dimension to what at first seems to be a non-enigmatic location: the basilica of Notre Dame de Marceille. There, from the 17th century onwards, the same people who were involved with the Compagnie (the Fouquet brothers, Nicolas Pavillon, Vincent de Paul, Jean-Jacques Olier, etc.), were involved in trying to lay their hands on this property. Why? On the surface, the church is like any other, but it has a hidden dimension, which has been successfully kept hidden for more than 500 years: underneath the church lies a secret underground complex, in which not only a monetary treasure was located, but which was also a pagan religious site, which was directly related to "the secret" – and which was once placed under the control of the Templars. After their demise, both the Compagnie and Saunière and his handlers tried to get hold of this underground complex, and its monetary and religious treasure – and of "the secret".

141 Pages. Paperback. 9.99 GBP / 14.95 USD

The Wanderings of the Grail *The Cathars, the search for the Grail and the discovery of Egyptian relics in the French Pyrenees*

André Douzet

In the 13th century, the Church came down against the Cathars, who had settled in the French Pyrenees. The Cathars practiced a belief in which "perfects" acted as priests that educated their followers in a specific system of believes and who aided the believers in "dying consciously", which was also at heart of ancient Egyptian belief systems. Both the Egyptians and the Cathars felt they had to "cheat" the cycle of reincarnation (the cycle of evil), and "ascend" to the world of light.

In the 20th century, both local amateur archaeologists and German Nazis such as Otto Rahn became interested in Catharism and sent investigators to the region. The Germans specifically were searching for the Holy Grail. They uncovered the ancient sanctuaries of the Cathars – often caves in perilous locations – and found Egyptian artefacts in them: statues of the Egyptian gods. Was the Grail perhaps a sacred Egyptian artefact?

One such researcher was Déodat Roché, nicknamed the Cathar Pope. As magistrate of Arques, he came across Egyptian relics scattered in his town which had been known as a major Cathar stronghold. He continued his researches further afield, in the heartland of the Cathars, where he learnt how others, including a priest named Glory, had found an ancient Egyptian statue in a cave, which was the centre of worship of the Cathars.
After World War II, these findings were quickly destined to be forgotten, because of the Nazi connection. Hence, the truth of the Cathar religion could once again not be made public. For the first time, all the key ingredients will be pieced together, and the enigmatic relationship between the Cathars and the Templars highlighted.
98 Pages. Paperback. 9.99 GBP / 14.95 USD

Nostradamus and the Lost Templar Legacy
Rudy Cambier

Rudy Cambier's decade long research and analysis of the verses of Nostradamus' 'prophecies' has shown that the language of those verses does not belong in the 16th Century, nor in Nostradamus' region of Provence. The language spoken in the verses belongs to the medieval times of the 14th Century, and the Belgian borders. The documents known as Nostradamus' prophecies were not written in ca. 1550 by the French 'visionary' Michel de Nostradame. Instead, they were composed between 1323 and 1328 by a Cistercian monk, Yves de Lessines, prior of the abbey of Cambron, on the border between France and Belgium. According to the author, these documents reveal the location of a Templar treasure. This key allowed Cambier to translate the 'prophecies'. But rather than being confronted with a series of cataclysms and revelations of future events, Cambier discovered a possible even more stunning secret. Yves de Lessines had waited for many years for someone called 'l'attendu', the expected one. This person was supposed to come to collect the safeguarded treasures of the Knights Templar, an organisation suppressed in 1307. But no-one came. Hence, the prior decided to impart the whereabouts and nature of the treasure in a most cryptic manner in verses. 220 years later, this document was stolen from a library by Nostradamus, who would make the enigmatic texts world famous, claiming they were 'prophecies'. The story, however, does not end here. The location identified in the documents and discovered by Cambier has since been shown to indeed contain what Yves de Lessines said they would contain: barrels of gold, silver and documents.
192 Pages. Paperback. GBP 11,99 / USD 17,95

191